SpringerBriefs in Applied Sciences and Technology

SpringerBriefs present concise summaries of cutting-edge research and practical applications across a wide spectrum of fields. Featuring compact volumes of 50 to 125 pages, the series covers a range of content from professional to academic.

Typical publications can be:

- A timely report of state-of-the art methods
- An introduction to or a manual for the application of mathematical or computer techniques
- A bridge between new research results, as published in journal articles
- A snapshot of a hot or emerging topic
- An in-depth case study
- A presentation of core concepts that students must understand in order to make independent contributions

SpringerBriefs are characterized by fast, global electronic dissemination, standard publishing contracts, standardized manuscript preparation and formatting guidelines, and expedited production schedules.

On the one hand, **SpringerBriefs in Applied Sciences and Technology** are devoted to the publication of fundamentals and applications within the different classical engineering disciplines as well as in interdisciplinary fields that recently emerged between these areas. On the other hand, as the boundary separating fundamental research and applied technology is more and more dissolving, this series is particularly open to trans-disciplinary topics between fundamental science and engineering.

Indexed by EI-Compendex, SCOPUS and Springerlink.

Marco Mancini

The Basalt Fiber—Material Design Art

 Springer

Marco Mancini
Department of Architecture
University of Ferrara
Ferrara, Italy

ISSN 2191-530X   ISSN 2191-5318 (electronic)
SpringerBriefs in Applied Sciences and Technology
ISBN 978-3-031-46101-9   ISBN 978-3-031-46102-6 (eBook)
https://doi.org/10.1007/978-3-031-46102-6

This Springer imprint is published by the registered company Springer Nature Switzerland AG
The registered company address is: Gewerbestrasse 11, 6330 Cham, Switzerland

Paper in this product is recyclable.

*I dedicate this book to those who,
more or less consciously,
have given me the motivation,
strength and impetus
to pursue this research project.*

To my roots, to my fruits.

Foreword by Mario Tozzi

I have always been fascinated by basalt rocks, ever since I discovered, as a boy, that this was the origin of the Roman *basolati* and *sampietrini* (flagstones and cobblestones) that form the streets of our capital and other cities, such as Naples and Catania. In truth, it was not quite like that: the rock of the *basolati* and *sampietrini* is improperly called basalt, which is in fact a complex mixture of rocks with different names. It is often leucitite, that is a rock poor in silica and rich in potassium, with large, milk-white, almost spherical crystals. Other times it is a tephrite, lava also poor in silica, but which does not develop such conspicuous leucite crystals; or a phonolite, still a similar rock, but which, when struck with a hammer, sounds characteristically (as the name implies). But let us say that the family is the same, and therefore, my approximation was justified, not least because, outside of the Eternal City, basalts in the strict sense of the word are often involved. Basalt is fascinating, then, but where does it come from?

Volcanism is one of the Earth's fundamental dynamic processes and is the evidence of the internal activity of the Earth 'machine,' as it continuously transfers heat and material (magma) from the interior to the surface. Volcanoes are also open windows to the Earth's interior: it is only thanks to volcanoes that we can directly know the first 300 km beneath our feet.

As can easily be seen by looking at a planisphere of the Earth's surface, volcanoes are not found scattered randomly on it, but are distributed only along a few narrow and elongated zones, nearly always on the edge of continents or in the middle of oceans. While those close to the continents are well-known volcanoes—such as those in Japan and the Americas—the others are not as well known, mainly because their activity occurs some thousand meters below the sea surface. Then there are other volcanoes that are not found in either of these 'regions,' but are apparently scattered within the continental and oceanic plates (and not at their margins). The different geographical locations correspond to different geological environments, and these diversities explain why volcanoes are not all the same as they might seem.

In this case, we are interested in the volcanoes of the oceanic ridges, which open directly on the depths of the crust, from which a very fluid and very hot (over 1000 °C/ 1832 °F) basaltic lava continually emerges, which is nothing else than new crust produced every day by the Earth. We are also interested in that minor volcanism dispersed within the lithospheric plates (i.e., not located at the margins alone) that is traced, by most scholars, to the action of hotspots; more than two hundred volcanoes like these are known on the Earth's surface, and some geologists believe that Etna in Sicily may also have had this type of origin. In all of these cases, basalt is the main lava emitted, making it the most represented on the planet.

Oceanic ridges make up the most important known eruptive apparatus, some 64,000 km long, that affects the entire Earth. From a physiographic point of view, they are massive submarine reliefs (hundreds of kilometers wide and up to 2000 meters high in relation to the abyssal plains) carved by a deep middle rift (Rift Valley), from which flowing basaltic lavas continually flow out. While ridges are the zones in which sectors of the lithosphere (plates) divide and move away from each other, rift valleys often—but not always—correspond to the zones in which the lithospheric plates collide (subduction zones).

It is worth remembering that what can be observed on the surface is only the outer and least important part of a volcano, even though it is the only one from which positive information can be obtained about the interior of the Earth's crust itself. In fact, the external apparatus can have very variable shapes, either a pointed cone, like some Japanese volcanoes, or very low and flattened, like the volcanoes of Iceland. The shape depends essentially on the type of products emitted: basaltic lavas will tend to give rise to low, soft-profile volcanoes, as can also be seen in Etna, while more acidic lavas will give rise to typical pointed shapes, such as Fujiyama.

The eruptions from which the basalt originates can also differ: 'Hawaiian'-type eruptions are very quiet: basaltic lava flows fluidly out of the conduits, and gases are released without any explosion. Lava fountains and lakes are characteristic of this type of eruption and are major tourist attractions on islands such as Hawaii or La Réunion. Eruptions of the 'Icelandic' type are just as quiet: large basaltic expansions are formed by the continuous outflow of lavas from very long cracks that open on rather flattened apparatuses, such as the Laki in Iceland. In the geological past, eruptions like these were very common and formed the large basalt expansions that can still be observed today in India (Dekkan) or South America (Rio Paraná).

The history of basalt is the very history of planet Earth; perhaps, for this reason, sapiens have often given this rock a special, even religious or sacred value. And they have always tried to use it, and not only to make roads. In Iceland, basalt is currently used to absorb excess carbon dioxide produced by the island's power plants. It is not surprising, therefore, that it is possible to make fibers from basalt with definitely interesting properties and with less environmental impact than traditional fibers. In this book, a new story is told of one of the world's oldest and most versatile natural materials—a story that could help us in the overall ecological reconversion that humanity is inevitably undergoing and a story that teaches us more than we can imagine.

Consiglio Nazionale delle Ricerche, Rome, Italy   Mario Tozzi

The Matter and The Invention

In the recent decades, a major thematic polarization has affected the field of design: the progressive dematerialization of artifacts and their digital transcription in the form of services have been balanced by an increasing interest in the physical texture and material quality of objects and environments. When they seemed to be disappearing, materials regained a pivotal role in magazines and in the realizations of architects and designers.

On the physical side of artifacts, this dichotomy has generated both an increase in research into new materials and the application of new processing techniques to traditional materials.

The return from the bit to the atom (as the Makers like to call it), the focus of design on materials, and their esthetic qualities, as a reaction to the flight into parallel universes generated by the digital revolution, must not be confused with a mere opposition between nature and artifice.

Almost 40 years ago, in his book *La materia dell'invenzione* (The Matter of Invention),[1] Ezio Manzini proposed an articulate mapping of the relationship between materials and the artificialization of human environment. On the relationship between the natural and the artificial and its cultural implications, Manzini observed that "the transformation of materials, manufacturing processes, and technological knowledge produces an artificial that questions its own traditional recognizability and the system of spatiotemporal relations we establish through it" (Manzini, 1986: 27). Therefore, faced with this loss of identity materials, with the ever-increasing difficulty of defining a material esthetically through a series of invariable categories, Manzini wrote, "the only way to describe it is to consider it as an operator endowed with performance: that is, to speak of the material not by defining 'what it is,' but by telling 'what it does'" (Manzini, 1986: 29).

Even stone materials, among the oldest ones used by man in the fields of architecture, art, and the production of tools and furnishings, have been constantly affected by processes of artificialization over the past thirty years.

---

[1] Manzini, E. (1986), *La materia dell'invenzione*, Milano, Arcadia.

The potential offered by numerically controlled machining 'by taking away' (as the work of Raffaello Galiotto so aptly demonstrates), or by the reduction of thickness in the production of upholstery elements, has expanded the use of these materials in architectural and interior design projects and particularly in the field of fashion.

However, even the reduction of a natural material such as stone to a mere information or visual code of positioning in the luxury market, as was already the case in Imperial Rome, is also in many respects a process of artificialization, in which the cultural datum prevails over the 'natural' one. This is as valid today for stone as it is for carbon fiber.

In the Department of Architecture at the University of Ferrara, the Material Design Laboratory, coordinated by Alfonso Acocella, has been active for many years. The laboratory is the publisher of the scientific journal *MD Journal*, which has devoted several issues to the theme of research on materials and stone materials in particular: #01 'Sensitive Enclosures'; #06 'Stone Design'; and #12 'Stone and Time.'

The laboratory has also supported the publication of several monographs dedicated to the themes of stone design, and some of its researchers were the promoters of the recent conference dedicated to the relationship between design and natural fibers, where some research dedicated to the themes of mineral fibers found space and moments of confrontation.

Bridging the gap between stone design and design with fibers, the theme that Marco Mancini explores in this volume fits well into this research path, for its willingness and ability to investigate a 'traditional' material in its contemporary applications and implications.

Within the vision that suggests defining the identity of a material not for what 'it is,' but for how it behaves, for what 'it does,' basalt fiber certainly represents an emblematic example, because it starts as a magmatic stone of eruptive origin and ends up spun into fiber and woven into the form of fabric, endowed with remarkable and precious physical–chemical characteristics suitable for multiple applications also in the field of design, from sport to automotive, from nautical to medical design.

In this account, Mancini takes us on a journey from the mythological suggestions of the eruptive material to its current uses in the field of design. It is an itinerary that, to use his own mythological references, goes from Vulcan to Arachne, from casting to weaving.

Obviously, as with other amorphous fibers spun from rocks or siliceous sands, basalt requires, in many of its applications, consolidation by means of hardening resins. This has been a problem for centuries for all natural fibers that, for reasons of 'shaping,' needed to be amalgamated with other substances: such as straw latticework with lime and vegetable fiber weavings with lacquer. Here, the biggest sustainability problem (like that of all resin-based composites) is linked to the material's end of life and concerns the possibility of separating the two materials, a process in which basalt seems to have fewer problems than glass or carbon fiber.

If it is true, as Frank Lloyd Wright argued, that "architecture is in the nature of materials," the question today is what, in the field of design, is the nature of materials that, like basalt, bring with them some important characteristics of their natural (geological) origin, but none of their esthetic features.

Wright was referring both to the nature of anisotropic 'natural' materials such as stone and wood and to that of isotropic artificial materials such as brick and glass: consider the experimental usage of glass that Wright made at the Johnson Wax factory in Racine, for whose use in tubular forms he also filed several patents.

Therefore, to discover the 'nature' of a material, to expand its possibilities of use, to verify its results in terms of esthetic outcome, the only possible path is the experimental one, that is, the proposal, the suggestion, the development of experiences, and their documentation in terms of research.

In this sense, Mancini's work represents a useful guide to the knowledge of the nature, production processes, and application possibilities of this material, whose introduction in many fields is relatively recent and whose potential in terms of experimentation is still largely unexplored, as demonstrated by the concluding part dedicated to case studies and teaching experiences.

If Bachelard's quotation placed by Mancini in the preface is valid, not only in terms of suggestion, namely that "one can study only what one has first dreamed about", this implies that the best research directions are those in which we not only use the most correct methodological approaches, but also glimpse possibilities for exploration that resonate with our sensibilities.

University of Ferrara, Italy   Dario Scodeller

Preface

One can study only what one has first dreamed about (Gaston Bachelard).[2]

Basalt is the most abundant effusive rock in the Earth's crust, a sustainable material sprung from the bowels of our planet with truly remarkable mechanical, physical, and chemical properties that are closely linked to its genesis, more decisively than other materials. Through the magma, fire becomes Earth by cooling in contact with air and water: this is how basalt is born. Inevitably, there is a symbolic reference to the four elements of the classical tradition, i.e., the roots (*rhizōmata*) of Empedocles, responsible together with Love and Strife for the union or separation (birth and death) of things, the same elements that would later be taken up again in the alchemical experience.

Volcanic phenomena, from which basalt is generated, have also featured in myths and legends since ancient times. In *The Psychoanalysis of Fire*, Gaston Bachelard argues that it is conceivable that "the *objective* attempt to produce fire by rubbing is suggested by entirely intimate experiences […]. The love act is the first scientific hypothesis about the objective reproduction of fire" (Bachelard, 1964: 23–24). This is a fascinating and unverifiable hypothesis; however, it is conceivable that knowledge, especially in archaic times, arose mainly from fascinations, only to become experience and scientific method in later centuries. Through myth, it was possible to narrate and describe natural phenomena that could not otherwise be explained. Vulcan for the Romans, and even prior to that Hephaestus for the Greeks, lord of metalworking, was the god of fire that produces and generates, a symbol of creation (but also of death, if we remember that Empedocles himself, the theorist of the four natural elements, according to legend died precisely because of fire, throwing himself into Mount Etna). In the myth of Prometheus, fire is a symbol of knowledge: he steals a 'seed' of fire to donate it to men: hence the complex that drives sons to know and learn more than their fathers, even performing sometimes revolutionary acts.

---

[2] Bachelard G. (2010), *L'intuizione dell'istante. La psicoanalisi del fuoco*, Bari, Dedalo (I ed. ital. 1973).

This publication arises, on the one hand, from the fascination with the 'mythical' nature of basalt, born from the power of Gaia (Mother Earth) and, on the other hand, from the awareness of its uniqueness: basalt is the only natural material that can be simultaneously sculpted, cast in molds, and suitable for generating textile-type fibers. These two instances have guided a path of cognitive analysis on the material, its characteristics, its production, and its uses, aimed at formulating new hypotheses for design research, mainly in the fields of art and product design.

Part I of the volume presents a cognitive study of the material. Chapter 1 reports on fiber properties (morphology, mechanical properties, thermal properties, corrosion resistance, and dielectric properties) in order to provide an indication of the material's performance and expectations in application. This is followed by an explanation of the production technologies (Chap. 2), which vary slightly depending on the type of fiber to be obtained (discontinuous or continuous); since in many ways the production process is similar to that of glass fiber, it was considered interesting to compare the two materials. Once the characteristics and production technologies are understood, we move on to describe the areas of application (Chap. 3) of an established type (excluding for the time being some experiments, dealt with in the second section of the volume). Part I closes with a reflection (Chap. 4) on the theme of sustainability and how this material, given its characteristics, can contribute to this closely topical issue, possibly also encouraging new production investments, of local scope, in line with environmental regulations and optimized in management thanks to the new frontiers of scientific research.

Part II begins with a reflection that explains how only through practical, applied research can the potential of a material be verified in areas more related to design and art (Chap. 5). Indeed, it is typical of these disciplines to take a physical, experimental approach, which passes through both intuition and error. When verified in this way, a design process of material research undoubtedly also brings positive results because the esthetic component of the product and the ethical dimension of the sustainable choice of material become complementary, strengthening the perception of the material and contributing to the tangible communication of its properties.

Thus, the course of research on the material carried out by the author of the text in collaboration with his students/artists of the courses in Design Culture, at the Academy of Fine Arts in Florence, Italy, is presented. The threads of research are grouped into four thematic areas: texture (Chap. 6), volume (Chap. 7), light (Chap. 8), and new proposals (Chap. 9); with the complicity of the innate curiosity that characterizes the artists, the various projects seek to overcome the limits of the consolidated use of the material, dissecting qualities and characteristics in depth and returning nontrivial proposals, worthy of reflection and evaluation for future areas of application of this natural material, in between art and design.

Cortona (AR), Italy   Marco Mancini

Acknowledgements I would like to thank Mario Tozzi for his opening foreword; Dario Scodeller for his fundamental advice, for his foreword, and important active contribution in the revision of the text; Francesco Rossi for providing me with the first important indications on the basalt fiber from which I started with my research work; Francesco Mollica and Marco Manfra for their valuable remarks in reference to some scientific contents; and Juri Ciani for his amazing photos and for his support.

I also want to remember with affection my friend and colleague Massimo Rossi, who laid the foundations for the achievement of the cultural project at the Academy of Fine Arts in Florence from which this publication originated.

Contents

About the Author

Marco Mancini is an architect, Ph.D. in design, professor and researcher in the fields of product design, innovation theory, technology for industrial design, focusing in particular on design for cultural heritage, for musical industry and events, and exhibition design for urban spaces; he has taught in bachelor's and master's degree courses at the University of Florence, University of Ferrara, Academy of Fine Arts in Florence, Academy of Fine Arts in Macerata, and Superior Institute for Artistic Industries (ISIA) in Pescara. He was a visiting professor at the School of Architecture at the Lisbon University (Portugal), Nanjing University of Aeronautics and Astronautics (China), and at the Dongguan University of Technology (China). He is an ordinary member of the Italian Design Society (SID) and the delegate of the National Council of Architects to the Technical Commission TC/33 "Products, processes, and systems for the building" of the Italian Standards Organization UNI.

He was an invited speaker at international conferences such as Cumulus Global Association of Art and Design Education and Research, European Academy of Design (EAD), European Research on Architecture and Urbanism (EURAU), and Information and Research on Reconstruction (i-Rec); he has published for editors such as Taylor & Francis, Springer Nature, Palgrave Macmillan, DIID, Franco Angeli, and Carocci.

He participates in inter-university research and development projects with European, national, and regional funds. He has designed installations and temporary setups for exhibitions of international relevance, including Pitti Immagine in Florence; he holds patents and design filings in the field of design for the music industry and in cultural heritage protection.

He is a jazz pianist, composer, and songwriter registered with SIAE under the pseudonym Mancinix.

Part I
The Material

Chapter 1
The Basalt Fiber

1.1 Basalt

There are many materials that can be totally or partially processed in liquid or semi-liquid form, then solidified in thermal processes: glass is heated to a high temperature and then blown or molded, but it is very difficult as well as risky to work it with a chisel to obtain sculptures; the ceramic is shaped or molded, then heated until it becomes a resistant product which, however, cannot be sculpted; plastic granules are heated and injected into molds, plastic sheets are thermoformed, but they again (whether with thermoset or thermoplastic matrix) cannot be subsequently sculpted; bronze, the material of choice for much of western sculptural art, once cast in molds and solidified can only be worked superficially; the cement, which can be molded and cast in variable shapes and sizes, once solidified can only be worked roughly.

On the other hand, materials that can be worked with the sculptor's tools (chisel, stone-cutter's chisel, rasp, bush-hammer…) either coarsely or finely, with varying degrees of precision and finish, cannot be melted: marble, wood, and limestone cannot be heated at high temperature and generate new products.

All this, however, is extraordinarily permitted for basalt, which can be, reversibly, heated to liquefy again and take on new forms.

With basalt, the loop of Empedocles' reasoning comes full circle: the four elements remain unchanged over time, generating all things through their mutual interaction.

Basalt is an effusive rock of volcanic origin (Fig. 1.1), consisting mainly of iron and magnesium silicates, calcic plagioclase, and pyroxenes, generated by the outflow of magma from the lithosphere. When the magma encounters the Earth's atmosphere or seawater, rapid cooling and pressure reduction arrest the crystallization process, giving the material a very compact, microcrystalline, porphyritic, or glassy paste structure (Fig. 1.2), and dark in color or even black (Caretto et al. 2017: 7). Basalt has a specific weight of almost 3 kg/dm^3 and can be extremely hard, with values between 5 and 9 on the Mohs scale. This igneous rock can be classified into two

© The Author(s), under exclusive license to Springer Nature Switzerland AG 2023   3
M. Mancini, *The Basalt Fiber—Material Design Art*,
SpringerBriefs in Applied Sciences and Technology,
https://doi.org/10.1007/978-3-031-46102-6_1

Fig. 1.1 Active lava flow, Kīlauea East Rift zone, Hawaii. Courtesy of Michael Grund

main groups, alkaline and calc-alkaline basalts, with silica contents ranging from 45 to 52% (Sheldon 1977: 18).

Basalt is a material that has been used for thousands of years for long-lasting sculptural works, suitable for both indoor (it does not release toxic substances) and outdoor use (it resists weathering). The Egyptians used it extensively for sculptural groups and to make sarcophagi to be kept inside pyramids, while examples of its outdoor use include the eighteenth-century elephant fountain in Catania, Italy. Its hardness and superior resistance to abrasion have been the basis of its widespread use as a paving stone for driveway or pedestrian road surfaces, or as a building material. Many urban landscapes are characterized by the chromatic and textural properties of this rock, such as the Rhineland area in western Germany, whose basalt buildings are photographed and collected in Arne Schmitt's book "Basalt" (2018).

When technological developments made it possible to create ad hoc furnaces, it was realized that it could be cast and poured into molds, so it was used for piping components or other elements in the industrial sector, especially because of its fire resistance characteristics (it comes from fire and is fireproof) and resistance to chemicals. A typical use of molten basalt is in the manufacture of rock wool, an excellent insulator and vibration-damping material. To obtain rock wool, stone is melted and then dropped while being blown at high speed: the glowing droplets are transformed into inhomogeneous fibers that adhere to a kind of carpet, covering it with rock wool.

Fig. 1.2 Basalt columns, Stuðlaberg, Reynisfjall Mountain, Iceland. Courtesy of Jennifer Boyer

There is an increasing use of basalt in the various fields of design: as a stone, it is used in claddings and tops for kitchens, for outdoor and indoor tiles, and for various objects. As a fiber, on the other hand, it is currently used in applications of high technological value, mainly in composites with resins, due to its remarkable mechanical, physical, and chemical properties. One of the motivations that guided the research paths presented in this text was the desire to verify the possibility of its application in the fiber form also as a visible material, attempting to propose its own esthetic value in order to extend its range of use, with a view to all-around sustainability.

1.2 Fibers from Basaltic Stones

The high availability, low cost, and uniformity in chemical composition (Fig. 1.3) make basalt an excellent raw material to produce technical fibers, better known as basalt fibers (Sheldon 1977: 18). But "though basalt stones are available in different compositions, only certain compositions and characteristics can be used for making the continuous filaments with a diameter range of 9–24 μm (Saravanan 2006)."

The technological development leading to industrially usable basalt fibers—which are therefore controllable in terms of quality, quantity, and size—has a relatively

Fig. 1.3 A kind of chemical
composition of basalt fiber.
Author's elaboration

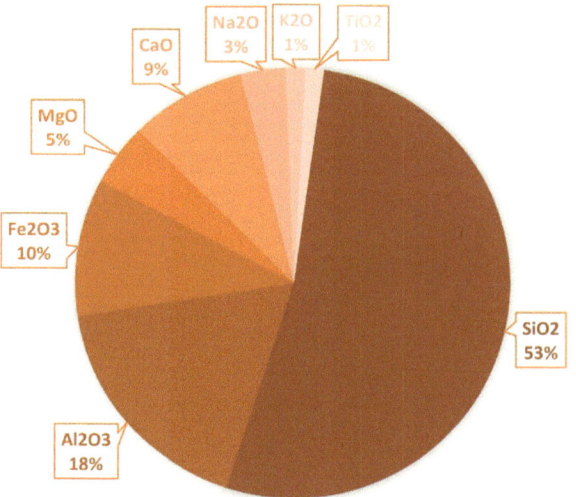

recent history. In 1923, the Frenchman Paul Dhé obtained a patent about the production of fiber from basalt rock, but there would not be any significant technological breakthroughs until the 1950s when the first research into producing fiber from basalt was developed in the Soviet Union (Levita 2008: 8), in the Moscow Research Institute of Glass and Plastic (Fiore et al. 2015). At the same time, research also took off in the United States, thanks in part to the presence of large basalt formations in the Northwest regions (Abdiev and Safarov 2022: 216); however, perhaps also due to political and industrial choices, research on new fiber-based composite materials in the United States focused mainly on glass fibers. The first production of basalt fiber, with still very complex and energy-intensive manufacture technologies, is reported to have taken place in 1985 in Ukraine, for military purposes. In the late 1990s, new-generation technologies were used to obtain fibers of controllable and homogeneous thickness, length, and properties, the production of which is currently located in Russia, Ukraine, and China. In many respects, the production technology of basalt fibers is similar to that of glass fibers; the ability to form a fiber is highly dependent on viscosity, which in turn has a strong dependence on temperature. For this reason, very tight temperature control of the melt material is necessary (Sheldon 1977: 18). The prevailing technology for obtaining fibers from basalt involves heating the chosen and selected material up to 1400–1500 °C/c. 2600–2700 °F for 24 h and then employing mechanical drawing systems using a resistance-heated rhodium–platinum bushing (Liu et al. 2022: 4).

This brief mention of the production technology, to which we will return more extensively in the course of the exposition, allows us to state that basalt fiber (Fig. 1.4) belongs to the category of natural fibers of mineral origin (Frassine et al. 2008: 14). In literature, the type of fibers is generally divided into the three categories natural, man-made, and synthetic (made by chemical processes). The natural fiber category includes fibers that originate from plants (cotton, linen, jute, sisal, hemp,

Fig. 1.4 Detail of basalt fiber fabrics. Courtesy Juri Ciani

etc.), animals (wool, silk, etc.), and minerals (glass, ceramics, etc.). In particular, only stones are used for the production of basalt fiber, unlike carbon fiber, which is generated from polymeric precursors (rayon, polyacrylonitrile, aromatic polyamides, phenolic resins) or from residues of oil or tar distillation (Crivelli Visconti et al. 2009: 34). The result is a considerable gain from the point of view of sustainability, given also the possibility of reuse and re-introduction, within the cycle, of the waste or fibers themselves.

However, it is in terms of performance that basalt fiber is an outstanding material.

1.2.1 Morphology

Basalt fiber is brown with golden highlights, and it has a smooth cylindrical appearance with a homogeneous surface; the homogeneity and uniformity are due to the fact that there is only one raw material, without the addition of other components—as for example in the production of glass fiber—and furthermore, during the manufacturing process no volatile substances are generated, that can significantly alter the appearance and properties of the fibers. During the drawing phase, it is possible to obtain more or less fine fibers by managing the speed and temperature: increasing these parameters results in thinner fibers, and decreasing them results in thicker ones (Parmar and Mankodi 2016: 44).

Table 1.1 Properties of basalt fiber in comparison with other fibers

Fiber type	Basal diameter (μm)	Density (g/cm³)	Price (USD/Kg)
Basalt	6–21	2.65–3.00	2.5–3.5
E-glass	6–21	2.55–2.62	0.75–1.2
S-glass	6–21	2.46–2.49	5–7
Carbon	5–15	1.78	30
Aramid	5–15	1.44	25

Fiber type	Tensile strength (MPa)	Modulus of elasticity (GPa)	Elongation at break (%)
Basalt	3000–4840	89–93.1	3.1
E-glass	3100–3800	72.5–75.5	4.7
S-glass	4590–4830	88–91	5.6
Carbon	3500–6000	230–600	1.5–2.0
Aramid	2900–3400	70–140	2.8–3.6

Fiber type	Maximum working temperature (°C)	Softening point (°C)	Thermal conductivity (W/m K)
Basalt	700	960	0.031–0.038
E-glass	380	850	0.034–0.040
S-glass	300	1056	0.034–0.040
Carbon	500	–	5–185

Sources: Abdiev and Safarov (2022), Chen et al. (2020), Landucci et al. (2006), Liu et al. (2022), Militky and Kovacic (1996)

1.2.2   Mechanical Properties

With regard to tensile strength, basalt fibers show similar values to E-glass fibers. Characterization studies on the material (Levita 2008: 16) have shown that the tensile strength of basalt fibers decreases with increasing length, predominantly due to the higher concentration of imperfections during manufacture compared to glass fibers. The strength performance of basalt fibers improves, compared to glass fibers, with increasing temperature. The elastic modulus of basalt fibers is comparable to that of glass fibers, higher compared to E-type but lower compared to S-type. Elongation at break values is higher compared to carbon fiber but lower compared to glass one (Table 1.1).

1.2.3   Thermal Properties

The softening point of basalt fiber is significantly higher than that of E- and S-type glass fiber and also higher than that of carbon fiber. The maximum working temperature is also significantly higher than the other fibers in the comparison (Table 1.1).

It is also important to emphasize the lowest working temperature of basalt fibers, $-260\,°C/-436\,°F$. These data attest to a theoretical operating range from -260 to $960\,°C/-436$ to $1760\,°F$, making this material suitable for very specific applications, e.g., for use in aerospace or as a high-performance insulator (Liu et al. 2022: 6). Experimental studies (Yang 2015) have measured the residual strength ratio of different types of basalt fiber fabric at different temperatures, showing that at $400\,°C/752\,°F$, the residual strength ratio is 88–90%, at $500\,°C/932\,°F$ is 65%, at $600\,°C/1112\,°F$ is 38.8%, and at $700\,°C/1292\,°F$ is 28.6%. Basalt fibers can withstand high temperatures and even an open flame; e.g., they withstand a hydrogen flame with an adiabatic flame temperature of $2060\,°C/3740\,°F$, absorbing a lot of heat (Landucci et al. 2006: 3).

1.2.4 Corrosion Resistance

The alkaline elements present in basalt, such as magnesium, titanium, sodium, and potassium, give the material an excellent resistance to alkaline corrosion. A comparative study was carried out between basalt fibers and glass fibers on salt and alkaline corrosion resistance; material samples, 15 μm in diameter, were subjected to aging tests in solutions of sodium chloride (NaCl) and sodium hydroxide (NaOH) for fifteen days, with tensile strength measured every five days. The results showed that basalt fiber offers better corrosion resistance (Liu et al. 2022: 7). The material characteristic can be further implemented by adding certain components to the raw stone or by acting downstream in the production process with coating substances. Their good corrosion resistance offers interesting uses for basalt fibers, e.g., to make blades for wind turbines, located outdoors, and often directly on the sea (Wang et al. 2019).

1.2.5 Dielectric Properties

As it is not a conductor, the dielectric (insulating) properties of basalt fiber allow wide margins for use in various industrial sectors: production of printed circuit boards in the electronics industry, insulation material for high- and low-voltage electrical equipment, and casings for antennas and radar radio devices.

References

Abdiev J, Safarov O (2022) Basalt fiber—basic (primary) concepts. Web Sci: Int Sci Res J 3(4):213–240

Caretto F, Laera A, Casciaro G (2017) Studio di un materiale ceramico innovativo destinato alla produzione di fibre di basalto. Rapporto tecnico ENEA-RT-2017-22. https://hdl.handle.net/20.500.12079/6785. Accessed 10 May 2023

Chen XF, Zhang YS, Huo HB, Wu ZS (2020) Study of high tensile strength of natural continuos basalt fibers. J Nat Fibers 17(2):214–222. https://doi.org/10.1080/15440478.2018.1477087

Crivelli Visconti I, Caprino G, Langella A (2009) Materiali compositi. Tecnologie, progettazione, applicazioni. Hoepli, Milano

Fiore V, Scalici T, Di Bella G, Valenza A (2015) A review on basalt fibre and its composites. Compos B Eng 74(1):74–94. https://doi.org/10.1016/j.compositesb.2014.12.034

Frassine R, Soldati MG, Rubertelli M (2008) Textile design. Materiali e tecnologie. Franco Angeli, Milano

Landucci G, Rossi F, Nicolella C, Zanelli S (2006) Materiali compositi in fibra di basalto per la protezione passiva di apparecchiature soggette a getti incendiati. http://conference.ing.unipi.it/vgr2006/archivio/Archivio/2006/Articoli/400193.pdf. Accessed 26 Jan 2023

Levita G (2008) Proprietà e caratterizzazione delle fibre di basalto. In: BASFA—International workshop on BASalt Fiber Application, Atti del convegno, Cecina, Polo Tecnologico della Magona, 2007. Collana Ricerca Trasferimento Innovazione. Regione Toscana, Firenze

Liu H, Yu Y, Liu Y, Zhang M, Li L, Ma L, Sun Y, Wang W (2022) A review on basalt fiber composites and their applications in clean energy sector and power grids. Polymers 14(12):2376. https://doi.org/10.3390/polym14122376

Militky J, Kovacic V (1996) Ultimate mechanical properties of basalt filaments. Text Res J 66(4):225–229. https://doi.org/10.1177/004051759606600407

Parmar S, Mankodi H (2016) Basalt fiber: newer fiber for FRP composites. Int J Emerg Technol Eng Res 4(7):43–45. https://ijeter.everscience.org/Manuscripts/Volume-4/Issue-7/Vol-4-issue-7-M-10.pdf. Accessed 11 May 2023

Saravanan D (2006) Spinning the rocks—Basalt fibers. J Inst Eng (India): Text Eng Div 86:39–45. http://basalt.today/images/Spinning-the-Rocks-Basalt-Fibres.pdf. Accessed 11 May 2023

Schmitt A (2018) Basalt. Spector Books, Leipzig

Sheldon GL (1977) Forming fibres from basalt rock. New application for a well-established process. Platin Met Rev 21(1):18–24

Wang X, Zhao X, Wu ZS (2019) Fatigue degradation and life prediction of basalt fiber-reinforced polymer composites after saltwater corrosion. Mater Des 163(5):107529. https://doi.org/10.1016/j.matdes.2018.12.001

Yang JH (2015) Surface treatment on basalt fiber and its composites performance evaluation. Master's Thesis, Donghua University, Shanghai, China

Chapter 2
The Production of Basalt Fiber

The possibility of manufacturing fibers directly from the basalt raw material is exploited for various applications: different production technologies allow basalt fibers to be obtained in two main categories: discrete fibers (rock wool, chopped, flakes, scales and powders) and continuous ones that can subsequently be woven into many commercial formats.

2.1 Discontinuous Fiber Production

The so-called discrete or discontinuous fibers, whose applications do not require exact dimensional control (length and diameter) of the individual fiber, are essentially used for their thermal and acoustic insulation, heat and flame resistance, and vibration-damping properties. The basalt rock extracted from the quarry is crushed, washed, and stored in silos or batching units from which the material passes into gas-heated furnaces. In the furnace, the rock is melted at 1450 °C/c. 2600 °F and then transmitted by a special machine to centrifugally driven heads from which the molten material exits and detaches; the fibers resulting from this process are approximately 60–100 mm long and 6–10 μm in diameter (Czigany 2005). This technology involves some production variants, in the way the melt material comes out and in the rotating machinery.

2.2 Continuous Fiber Production

The continuous fibers are produced from a single raw material, basalt rock, which is processed by machines of a more advanced type than those used in the production of discontinuous fibers. These machines are arranged in a single production line or set

© The Author(s), under exclusive license to Springer Nature Switzerland AG 2023   11
M. Mancini, *The Basalt Fiber—Material Design Art*,
SpringerBriefs in Applied Sciences and Technology,
https://doi.org/10.1007/978-3-031-46102-6_2

up in modular units. Compared to the production of discontinuous fiber, the process of transforming basalt into continuous fiber requires more control, right from the first preparatory stages. Crushing must generate stones with a diameter between 5 and 40 mm; this step is followed by magnetic separation to discern metal parts, washing, and drying in natural air or in special dryers (Abdiev and Safarov 2022: 222).

The crushed basalt rocks are then conveyed into the furnace, a step which is normally divided into two stages, with separate heating and control systems; greater care and precision are required especially in the second heating stage, at the end of which the liquid material comes out of platinum–rhodium alloy bushings, which, by means of resistance heating, further regulate the temperature and consequently the viscosity of the fiber. The next step is the elongation and winding of the filament, with automatic speed control: during this process, the dimensional properties of the fiber can be altered by precisely calibrating the diameter. A sizing agent, that is a substance that imparts special properties to yarns, is also applied at these stages. A scheme of the basalt fiber production is shown in Fig. 2.1.

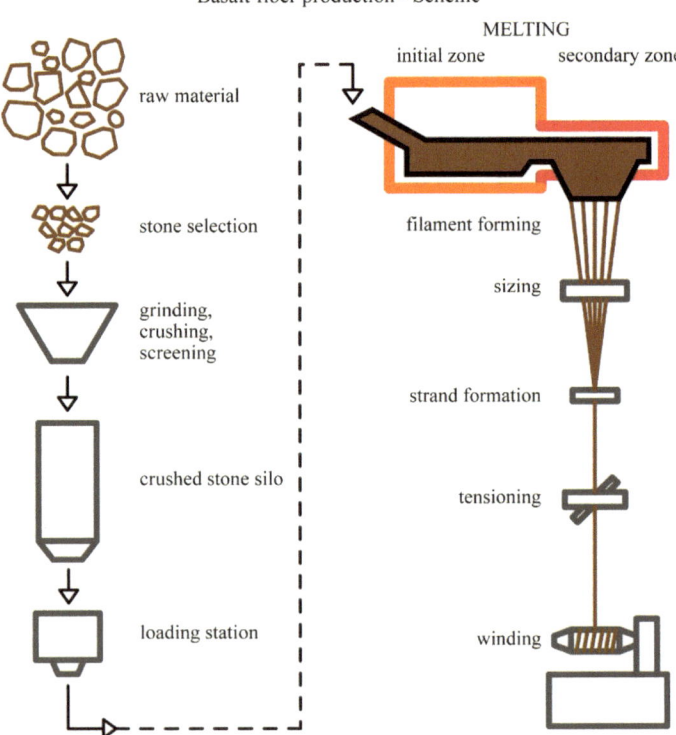

Fig. 2.1 Continuous basalt fiber production scheme. Author's elaboration

2.3 Basalt vs. Glass: a Comparison

Although the production technology for basalt fibers is conceptually similar to the one for glass fibers, it is important to emphasize some fundamental differences.

In the preparatory stages for processing the raw material, basalt rocks can be stored in the open air, without any special protection systems, with advantages in economic terms; in the furnace, while for glass the different raw materials (silica, boron oxide, aluminum, etc.) have to be dosed beforehand, that manufacturing complication is not present in the case of basalt fibers, as the enrichment of the raw material has already been done by the volcanoes from which it is generated. So there is only one furnace feed line, with savings in installation and management costs for the production system. However, in basalt, it is difficult to precisely control the chemical composition and purity, as these parameters depend on where the basalt was generated.

Although both are silicates, once melted and cooled, glass form a non-crystalline solid, while basalt presents a crystalline structure that varies depending on the specific conditions of the lava flow (e.g., cooling rate) and thus also on the geographical location of the quarry (Abdiev and Safarov 2022: 222). However, it has been verified (Novitskii and Sudakov 2004) that the mechanical properties of continuous basalt fibers depend only slightly on variations in the chemical composition of the raw material, whereas the drawing temperature, the time of homogenization of the mass, and the diameter of the filament have a decisive influence. With the same chemical composition of the basalt, increasing the drawing temperature of the fibers increases their strength and modulus of elasticity.

Another difference in the properties of the two materials requires a different processing technology: basalt, being opaque, absorbs infrared radiation, while molten glass is more or less transparent to it; in the production of fiberglass, this allows the material to be heated with gas burners, obtaining a homogenous temperature throughout the mass. By using this system with basalt, on the other hand, there is a significant variation in the thermal gradient depending on the distance from the heat source, so many hours are required to obtain a homogeneous temperature throughout the mass (Abdiev and Safarov 2022: 223), which also increases energy costs. To attempt to limit this problem, some manufacturers have developed heating systems using electrodes immersed in the material.

Fiberglass is a truly, widely used composite material in so many areas of production, from the most artisanal to highly industrialized. The chances that basalt fiber has to compete with this material also stem from extensive studies and comparisons in order to test not only its properties but especially its strength under certain conditions (Deak and Czigány 2009; Wei et al. 2010).

2.4   Surface Treatments

In composite materials, currently the main field of application experienced for basalt fiber, the fiber and the matrix play the role of support and load transfer, respectively; in this reciprocal collaboration, it is essential to verify and possibly improve the characteristics of the interface, i.e., the small surface portion where the matrix and fiber come into contact, which is usually a few microns or even a few nanometers. A good interface bond favors the transfer of an external load between the fiber and the matrix, preventing the concentration of stresses and effectively inhibiting the further expansion of any cracks. Conversely, when the interface bond is poor, the external load cannot be transferred well into the composite, resulting in a concentration of stresses at the defect with a consequent serious risk of damage to the composite. The surface of the basalt fiber is generally smooth and inert and does not offer optimal adhesion with the resin matrix, hence the need to improve the interface surface in order to optimally utilize the material characteristics. A study about this topic was developed by Greco et al. (2014).

The surface modification of basalt fibers is mainly of two types: physical and chemical. The physical modification affects surface roughness, specific surface area, and other microstructural characteristics that concern the mechanics of adhesion with the resin matrix.

The chemical modification occurs by introducing chemical functional groups with the aim of improving the bond strength of the interface. Currently, modification technologies mainly involve acid–base etching, surface coating as sizing, plasma treatment, treatment with coupling agents, treatment with nanomaterials, and treatment with sizing agents (Liu et al. 2022: 9). In the production of continuous basalt fibers, silane-based sizing liquid is predominantly used, which gives the filaments lubrication, greater integrity, and compatibility with any resins used.

References

Abdiev J, Safarov O (2022) Basalt fiber—basic (primary) concepts. Web Sci: Int Sci Res J 3(4):213–240

Czigany T (2005) Discontinuous basalt fiber-reinforced hybrid composites. In: Friedrich K, Fakirov S, Zhang Z (eds) Polymer composites: from nano- to macro-scale. Springer Science, Boston (MA), pp 309–28. https://doi.org/10.1007/0-387-26213-X_17

Deák T, Czigány T (2009) Chemical composition and mechanical properties of basalt and glass fibers: a comparison. Text Res J 79(7):645–651. https://doi.org/10.1177/0040517508095597

Greco A, Maffezzoli A, Casciaro G, Caretto F (2014) Mechanical properties of basalt fibers and their adhesion to polypropylene matrices. Compos B Eng 67:233–238. https://doi.org/10.1016/j.compositesb.2014.07.020

Liu H, Yu Y, Liu Y, Zhang M, Li L, Ma L, Sun Y, Wang W (2022) A review on basalt fiber composites and their applications in clean energy sector and power grids. Polymers 14(12):2376. https://doi.org/10.3390/polym14122376

Novitskii AG, Sudakov VV (2004) An unwoven basalt-fiber material for the encasing of fibrous insulation: an alternative to glass cloth. Refract Ind Ceram 45:239–241. https://doi.org/10.1023/B:REFR.0000046504.53798.af

Wei B, Cao H, Song S (2010) Tensile behavior contrast of basalt and glass fibers after chemical treatment. Mater Des 31(9):4244–4250. https://doi.org/10.1016/j.matdes.2010.04.009

Chapter 3
Uses and Fields of Application

Due to its characteristics, basalt fiber has numerous fields of application, some of which have been developed even recently. It is therefore appropriate to distinguish applications relating to fiber in discrete form from those that exploit production in continuous form.

3.1 Discontinuous Fiber Applications

The fiber in discontinuous form, in different weights and fractions—flakes, chopped (Fig. 3.1), scales (Fig. 3.2), and powders (Fig. 3.3)—is used essentially as an additive or reinforcement, capable of increasing product performance. In relation to its heat and flame resistance, it is used in products intended for a high-temperature environment such as refractory production or as an additive for non-combustible coatings. Its inertness, resistance to chemicals, and mechanical strength are exploited in the reinforcement of concrete (Adesina 2021), in the constituent mixture of fillers for epoxy components, e.g., in the automotive industry, or in the composition of dry plasters for repairs. In particular, there are advantages with regard to structural strength: in cements with basalt fibers, the tensile strength and elastic modulus are increased by 15–20% compared to those of traditional cements. To implement fiber-matrix adhesion, these fibers must receive a special surface treatment (sizing) based on inorganic salts (Landucci 2008: 27). Abrasion resistance encourages the use of basalt fiber, in composition with other substances, in many areas of mechanics: sliding bearings, plastic-based reinforced wheels, reinforced rubber for bogies, machines, and mechanisms, aluminum alloy reinforcements, and tool components. The combination of abrasion and corrosion resistance is exploited in the production of surface coatings, anticorrosive coatings in the marine environment, and self-leveling floors on epoxy or polyester basis (Fiore et al. 2011).

M. Mancini, *The Basalt Fiber—Material Design Art*,
SpringerBriefs in Applied Sciences and Technology,
https://doi.org/10.1007/978-3-031-46102-6_3

Fig. 3.1 Basalt fiber chopped. Courtesy by Rockfiber

Fig. 3.2 Basalt fiber scales. Courtesy by Rockfiber

The fiber produced in the form of wool is mainly used for its thermal and acoustic insulation properties and as an antivibration device; it is available in the form of carpets, sheets, and boards; the fields of application vary from construction to industry. It is used for example as a lining for engine compartments (in such applications, basalt resists 150 °C/c. 300 °F more than glass fiber), as protection for

Fig. 3.3 Basalt fiber powder. Courtesy by Rockfiber

household appliances, and as acoustic insulation in recording studios. It is also used as a filtration system. Commercial formats also include honeycombed basalt fiber panels, which are very light and insulating (Landucci 2008: 31).

3.2 Continuous Fiber Applications

The industrial production of fiber in continuous textile form has paved the way for the use of basalt in the field of composite materials, traditionally occupied mainly by glass and carbon fibers, in addition to aramid fibers. The composite comes out of a reinforcing material and a resin, typically epoxy, polyurethane, or polyester; the advantage is that lightweight, even complex-shaped, high-performance products can be obtained.

The roving, which is a roll of continuous filament, is the first subproduct generated by the transformation of basalt from rock into continuous fiber. Roving and yarn—shape variants (Fig. 3.4)—are used either directly or by means of preliminary textile processing.

Fig. 3.4 Basalt fiber in
continuous textile form,
wrapped in roving and yarn.
Courtesy by Basaltex

3.2.1 Direct Application

The technologies for using basalt fiber directly from continuous filament yarn essentially consist of filament winding and pultrusion. In the filament winding, the continuous fiber is taken directly from the roll and impregnated in a resin bath; before the resin is hardened, the impregnated fiber is tensioned and simultaneously wound around a rotating component—the mandrel—that gives the shape to the final product; after the total winding and once the desired shape is obtained (closed or open at the ends), the resin hardening process is completed by passages at high temperatures. The mandrel that gave the composite its shape can remain as an integral part of the structure, can be removed, or can be dissolved if soluble. Filament winding is mainly used to make tanks designed to withstand high pressures and, in the case of basalt fiber, also high temperatures and corrosion. The advantage of using this technique lies in the speed of execution, in the elimination of joints between the various components, and in the possibility of obtaining products with very high performance and at the same time lightweight, particularly suitable for extreme environments such as the space (Di Felice 2019).

The pultrusion, also known as pull forming, is a special extrusion process that allows the production of axially developed composite products such as profiles or bars. The initial procedure is the same as for filament winding, that is, there is the continuous drawing of the fiber from the roll and a subsequent impregnation with resin; in this case, however, the impregnated fibers are compacted and passed through the machinery that will give them their final section, before proceeding to the heating and hardening of the resin. Basalt fiber pultruded bars are used as reinforcement in concrete, replacing traditional steel. This field of application is very promising: a number of studies (e.g., Abdiev and Safarov 2022) reveal that reinforcement with basalt bars is approximately 89% lighter than with steel. Furthermore, basalt has the same coefficient of thermal expansion as concrete and is naturally resistant to corrosion, which has always been one of the greatest risk factors for the durability and stability of reinforced concrete. These characteristics are especially exploited in environments such as oil and gas platforms, port facilities, and buildings where high-performance resistance to chemicals is required.

3.2.2 Textile Applications

The continuous fiber roving that comes directly out of the production chain can then undergo other processes aimed at obtaining products such as nets or mesh or fabrics in the form of cloths or canvases with very different commercial names depending on the manufacturing companies (basalt-woven textile, basalt-knitted fabric, plain basalt-woven fabric, braided basalt fabric, etc.). Important characteristics are "the possibility of obtaining multi-axiality to confer specific polytropic structural characteristics or the possibility of having several layers of different types together, as occurs in the case of common fiberglass (woven + mat, woven + chopped, woven + nonwoven, etc.)" (Landucci 2008: 19). The fabric can either be distributed in rolls, sheets, or processed to form the so-called socks or scarves for coating.

Predominantly, the main field of use is in the production of composites, in combination with all types of resins (given the characteristic compatibility, guaranteed by the manufacturers); in some cases, when the material has to resist direct sources of heat or flame, it is marketed in the form of "socks" or "scarves" to cover, for example, pipes or structural elements, in this case without the use of resins. The different types of commercial products allow basalt fiber textiles to be used in a wide variety of areas (Tavadi et al. 2021: 54), as it is represented in Table 3.1.

In the construction industry, basalt fiber, already described as concrete reinforcement—when in discrete form (Parinya and Wichit 2018), as rebar (in the form of pultruded bars), as insulation (rock wool), is increasingly used as structural reinforcement (De Domenico et al. 2022), replacing carbon fiber, due to its better cost-effectiveness and great compatibility with the increasingly high-performance chemical products used in consolidation (Kumbhar 2015). A typical use is the encircling of pillars with more or less dense meshes (Fig. 3.5) in order to reinforce their resistance to compression; an increasing use is the consolidation of vaults with basalt fiber meshes placed on the extrados and then covered with plaster or resin of various kinds. With the aim of increasing performance, basalt fiber is also used in the construction of dams, retaining structures for embankments and protection against accidental rockfall, as reinforcement for pedestrian or vehicular sub-bases, and as reinforcement in asphalt (Morova 2013).

Basalt in the form of net or fabric is also widely used in the sports industry: skateboards (Figs. 3.6a, b) or surfboards are made from different layers of material, including basalt fiber fabric, which provides strength, elasticity, and vibration damping; for these characteristics, it is also used in the manufacture of table tennis paddles (Fig. 3.7) and ski and snowboards. Basalt fiber can also be used instead of carbon fiber to make structural components, such as bicycle frames.

Table 3.1 Main industrial applications of basalt fiber. Author's elaboration

Aerospace industry	High-performance components Sandwich panels Cladding components
Automotive industry	Fiber-reinforced structural components Sound insulation Vibration absorbing elements Components for headliner Exhaust systems Brake pads Fairings for cars and motorbikes
Sports equipment	Skateboards Ski and snowboards Surf boards and components Bicycles frames Table tennis paddles
Construction industry	Concrete reinforcement Vault consolidation Pillars reinforcement Insulation panels Dam buildings Rock falls protection Asphalt reinforcement
Maritime industry	Boat hulls Structural sandwich panels Insulation panels Nets for underwater activities
Other areas	High-pressure gas tanks Fire-resistant coatings for pipes Fireproof textile industry applications Components for wind turbines Components for military equipment Gaskets for various uses Filtration components

The automotive industry has realized that the potential of basalt fiber is perfectly compatible with the requirements of performance, insulation, resistance to heat, corrosion, and abrasion (Shoaib et al. 2022); in the automotive field, in addition to the already-mentioned applications of fiber in wool form, basalt fiber in fabric and composite form is used for structural components (Figs. 3.8a, b) such as some car roofs or body parts, in the realization of exhaust systems or components for braking systems, as acoustic insulation and vibration reduction, in composite with resins for aerodynamic elements. In addition to its performances, one of the reasons for choosing this material is also in its sustainability (maybe with positive communicative effects in terms of marketing): "as a naturally occurring material, basalt fiber is inherently more recyclable than other reinforcing fibers, a factor that automotive and other industries take into consideration" (Mason 2020).

Fig. 3.5 Basalt fiber nets, meshes, and geogrids. Courtesy by Basaltex

The chance to make high-performance composites has allowed the material to be appreciated for its characteristics in the nautical, aeronautical, and aerospace industries, both to make structural and cladding components, even in sandwich-type layered elements. In Italy, in 2012, Cantiere Tripesce and other Livorno-based nautical SMEs with co-financing from the Regione Toscana realized Tripesce 29B (2012), the first boat with a hull made entirely of basalt fiber, in order to propose a more sustainable and high-performance alternative to traditional fiberglass; in applications in the marine environment, basalt fiber has the advantage of resisting corrosion even when the composite's coating layer (gel coat) eventually becomes damaged, a frequent problem when using glass fiber. Sustainability, mechanical and weather resistance are key factors for the use of basalt fiber also in the construction of wind blade components.

Fig. 3.6
a Loadedboard—Tarab II.
Courtesy by Loadedboard
USA.
b Loadedboard—Tarab II.
Detail of the back surface.
Courtesy by Loadedboard
USA. The US company
Loadedboards (2023) has
been using basalt fiber since
2015 as a framework for
some lines of skateboards.
This fiber, in a biax and twill
weave, is applied to the two
sides of a central bamboo
core and then covered with
adhesive tape, cork, or
bamboo veneer. The use of
basalt fiber is advantageous
because it allows to obtain
the right compromise
between the stiffness of the
glass fiber and that of the
carbon fiber; moreover, it has
a high damping effect,
optimal for dancing, and
freeride/downhill-type
skateboards

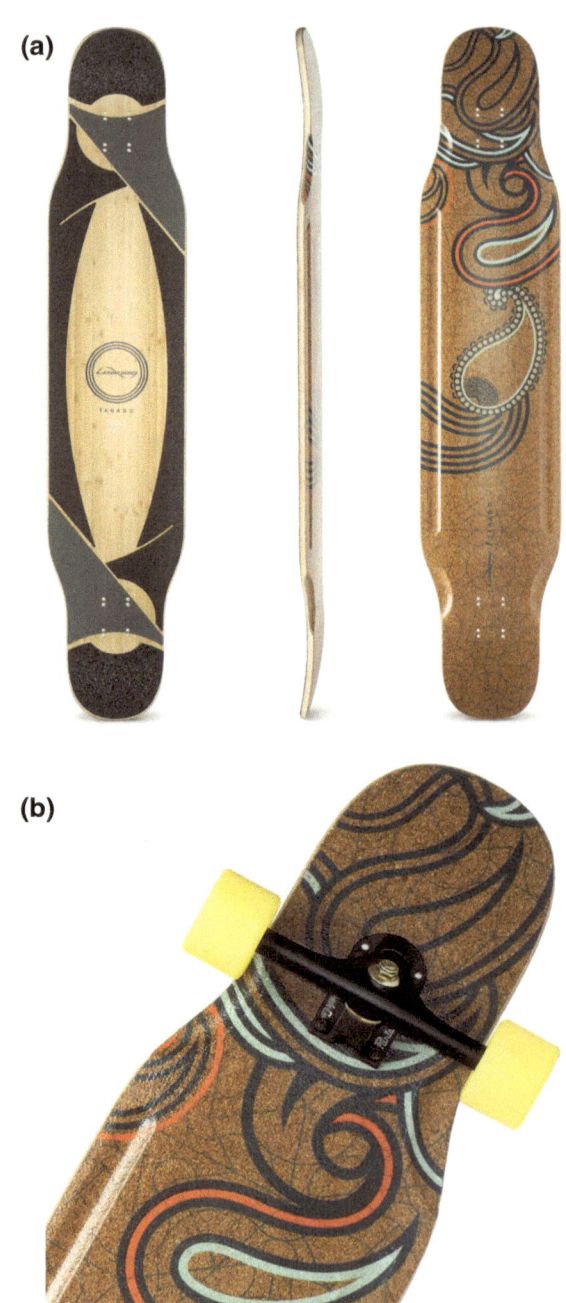

(a)

(b)

Fig. 3.7 Soulspin (2022) table tennis blade. This table tennis paddle is made of layers of different woods with basalt fiber layers in between. Courtesy by Soulspin table tennis blades

With regard to personal protection, basalt fiber composites are nowadays widespread in many fields such as in the health and orthopedic ones (braces and external prostheses) where, in addition to the strength, flexibility, and lightness of the material, non-toxicity and absence of harmful emissions are also important; basalt fiber composites are used in protection systems such as bulletproof vests instead of Kevlar; Landucci (2008: 38) cites a prototype of a basalt fiber fireproof suit that can be used in hazardous and extreme conditions with temperatures from − 260 to 750 °C/− 436 °F to c. 1380 °F.

Fig. 3.8 a Peugeot Exalt (2014). Courtesy by Stellantis. Based on a superior idea of sustainability, this car—presented in 2014—also uses basalt fiber as a structural composite. **b** Peugeot Exalt (2014). Courtesy by Stellantis. Detail of basalt fiber composite, with a natural varnish treatment, visible at the door entrance and low chassis areas

References

Abdiev J, Safarov O (2022) Basalt fiber—basic (primary) concepts. Web Sci: Int Sci Res J 3(4):213–240

Adesina A (2021) Performance of cementitious composites reinforced with chopped basalt fibres—an overview. Constr Build Mater 266(A). https://doi.org/10.1016/j.conbuildmat.2020.120970

De Domenico D, Maugeri N, Longo P, Ricciardi G, Gullì G (2022) Clevis-grip tensile tests on basalt, carbon and steel FRCM systems realized with customized cement-based matrices. J Compos Sci 6(9):275. https://doi.org/10.3390/jcs6090275

Di Felice A (2019) Filament winding: una nuova frontiera per la produzione di compositi. AerospaceCuE. https://aerospacecue.it/filament-winding-una-nuova-frontiera-per-la-produzione-di-compositi/12055/. Accessed 3 Feb 2023

Fiore V, Di Bella G, Valenza A (2011) Glass-basalt/epoxy hybrid composites for marine applications. Mater Des 32(4):2091–2099. https://doi.org/10.1016/j.matdes.2010.11.043

Kumbhar V (2015) An overview: Basalt rock fibers—new construction material. Acta Eng Int 2(1):11–18. https://www.researchgate.net/publication/302987042_An_overview_basalt_rock_fibres-new_construction_material. Accessed 10 Aug 2023

Landucci G (2008) Applicazioni delle fibre di basalto. In: BASFA—international workshop on BASalt fiber application, Atti del convegno, Cecina, Polo Tecnologico della Magona, 2007. Collana Ricerca Trasferimento Innovazione. Regione Toscana, Firenze

Loadedboard (2023). https://loadedboards.com/. Accessed 31 Jul 2023

Mason K (2020) The still-promised potential of basalt fiber composites. Compos World. https://www.compositesworld.com/articles/the-still-promised-potential-of-basalt-fiber-composites. Accessed 6 Feb 2023

Morova N (2013) Investigation of usability of basalt fibers in hot mix asphalt concrete. Constr Build Mater 47:175–180. https://doi.org/10.1016/j.conbuildmat.2013.04.048

Parinya C, Wichit P (2018) Feasibility study of using basalt fibers as the reinforcement phase in fiber-cement products. Key Eng Mater 766:252–257. https://doi.org/10.4028/www.scientific.net/KEM.766

Peugeot Exhalt (2014). https://www.media.stellantis.com/fr-fr/peugeot/exalt. Accessed 31 Jul 2023

Shoaib M, Jamshaid H, Alshareef M, Alharthi FA, Ali M, Waqas M (2022) Exploring the potential of alternate inorganic fibers for automotive composites. Polymers 14(22):4946. https://doi.org/10.3390/polym14224946

Soulspin (2022) https://soulspin.de/en/building-table-tennis-blades/table-tennis-blades-with-basalt-fibre. Accessed 31 Jul 2023

Tavadi AR, Naik Y, Kumaresan K, Jamadar NI, Rajaravi C (2021) Basalt fiber and its composite manufacturing and applications: an overview. Int J Eng Sci Technol 13(4):50–56. https://doi.org/10.4314/ijest.v13i4.6

Tripesce 29B (2012) la barca vulcanica costruita in fibra di basalto. https://www.mondobarcamarket.it/tripesce-29b/. Accessed 31 Jul 2023

Chapter 4
Sustainability

The basalt fiber, as mentioned, is produced from a single raw material, the natural basalt mineral. During the production process, unless it is necessary to obtain special characteristics, no other products such as chemical additives, solvents, or pigments are added; furthermore, at the end of the production process, there is no degradation, that is, the fiber has the same composition as the starting mineral, without the emission of harmful substances such as boron or other alkaline metal oxides. It is a totally inert material, with no reactions in contact with air or water; no chemical reactions have been observed in contact with substances; it is non-combustible and explosion-proof. In compliance with Regulation (EC) 1907/2006 REACH & (EU) No. 2015/830, continuous basalt filaments, although physically cut to predetermined lengths, are not classified as hazardous. With regard to the disposal of the material at the end of its life cycle, basalt fabric residues can be considered special non-hazardous waste (definition from Directive 2008/98/EC). As such, they can be deposited in approved landfills (small quantities can be disposed of together with municipal waste).

What is stated above refers to basalt fiber used in its natural state. A different approach is therefore necessary in the case of fiber used in composites with resins of various types, that is, in the majority of cases in which this material is used. The possibility of obtaining lightweight, very strong, and high-performance components has driven research into composites (Agarval and Broutman 1990) based on glass fiber, carbon fiber, or other fibers (Cooke 1991) including basalt ones until the development of numerous products used in very different sectors. Several decades after the first uses of composite products, the problem of the disposal of such fiber and resin products is increasingly emerging. To give an example, the aeronautics industry will be deeply involved in the problem of composite management: over the next decade, in fact, many of the components of commercial aircraft produced during the 1990s will reach the end of their lives. According to an estimate (Caretto 2020), on every aircraft at least 20% of the components, i.e., about 2 tons in weight, are made of composites; even if we limit the survey to the civil aeronautical sector, these are enormous quantities worldwide, of which only a fraction is reused, while almost

M. Mancini, *The Basalt Fiber—Material Design Art*,
SpringerBriefs in Applied Sciences and Technology,
https://doi.org/10.1007/978-3-031-46102-6_4

all are destined for disposal in landfills or incineration. Another area involved in the end-of-life management of composites concerns the production of energy from renewable sources, whose plants, even if intended for such sustainable production, also must be dismantled. Referring, for example, to the huge wind turbines, currently made almost exclusively of glass fiber, a report (Liu and Barlow 2017) predicts that about 43 million tons of waste from the decommissioning of current wind turbines will have to be managed worldwide by 2050.

The end-of-life management of composite products, therefore, follows different paths. Repairs or replacements can sometimes be carried out to extend the product's life. When the product is part of a larger artifact that needs to be replaced, then it is possible to recover individual parts and adapt them to new uses, keeping the characteristics of the composite unchanged but, for example, downsizing it as in the case of recovering portions of wind turbine blades for new uses in the manufacture of bridges or buses (Arussi et al. 2023: 12). An interesting case of recycling is that carried out by Gees Recycling, an Italian company that uses fiber-reinforced thermoset matrix composites to generate panels with different densities and recipes, which can be reused in place of MDF panels (as they can be processed with the most common joinery techniques: drilling, milling, screwing, laminating, painting, gluing) but more flexible in the choice of performance, featuring water-repellent properties, without formaldehyde, resistant to bacterial proliferation in humid environments, and with dimensional stability, unlike the latter. Products made of basalt fiber composites can enter this process and be reused, also thanks to basalt's mechanical resistance and compatibility with other materials.

When the composite product has lost its strength characteristics and shows more or less extensive and visible degradation phenomena, it is no longer repairable or recyclable for other uses. At this point, it is possible to refer to the product's end of life. The management of the material's decommissioning involves different stages depending above all on the type of matrix used. In the case of thermosetting resins, which represent the highest market share among fiber-reinforced composite products, most of the decommissioning processes generate a decay in performance: mechanical treatments such as mechanical grinding (mechanical abrasion), chemical treatments such as solvolysis (scission through a solvent), and thermochemical treatments such as pyrolysis (application of heat without an oxidizing agent) only allow the obtainment of additives/fillers to be reused in the composite cycle or for other uses. The last stage remains landfilling or incineration. As far as incineration is concerned, it is important to emphasize that the very high melting point of basalt—1400 °C/2552 °F—makes it theoretically possible to work at much lower temperatures in order to incinerate only the resin component while leaving the basalt fiber intact for subsequent reuse. Fiberglass, on the other hand, is difficult to handle in incinerators because the glass melts, thus creating a mass that adheres to the structures, which is difficult to remove.

4.1 The Future Scenario

The growing attention to issues related to the impact of human actions on the environment has led to a gradual awareness of sustainability. After initial attempts to come up with sustainable products through best practices and guidelines, the European Commission finally adopted the Ecodesign for Sustainable Products Regulation (2023) ESPR proposal in 2022, which aims to make almost all products placed on the EU market sustainable by law. Some important parameters to be considered will be:

- product durability
- the possibility of reuse, repair, updating
- the presence of substances that inhibit circularity
- energy and resource efficiency
- recycled content
- the possibility of remanufacturing and recycling
- carbon and environmental footprint
- information requirements.

Each product released on the market will be equipped with a "Digital Product Passport" that will provide information on environmental sustainability, helping consumers and businesses to make informed choices when purchasing, facilitating repairs and recycling, and improving transparency on the life cycle impacts of products on the environment. The product passport should also help public authorities to perform better checks and controls. This is a significant and indispensable step forward, which, in turn, will have implications for the search for innovation based on technological development.

In order to try to improve the efficiency of the material recycling phase, new methods are being developed for the composite materials sector, such as the fluidized bed or gasification, a thermal recovery process similar to pyrolysis but more effective in the case of mixed materials in which there are foams or painted surfaces in addition to the composite, or fragmentation with high voltage pulses, which, compared to grinding, generates better quality products because the fibers are kept more intact and longer (Assocompositi 2022).

A very promising approach is offered by research into new types of resins (Torres et al. 2013), primarily thermoplastic ones, which are easier to separate or dismantle or have a lower environmental footprint. New thermoplastic resins that can be recycled by depolymerization or dissolution are being studied (Cardone 2022); these new formulations allow the separation of the fiber from the resin in the composite, with considerable advantages in terms of environmental and economic sustainability.

In the case of basalt fibers, the natural sustainability of the material and its essential properties can play a crucial role in sustainable development. For example, the company Anisoprint (Luxembourg) produces reels of coextruded basalt fiber monofilaments, mixed in the same extruder with matrices of various kinds (such as PLA or PETG) with considerable advantages in terms of strength, stiffness, and lightness (Anisoprint 2023).

Frontier research, closely linked to current space exploration programs, concerns the possibility of producing semi-finished products directly on the Moon, from the local basalt in which our satellite is extremely rich. Such installations, built on site, could allow the construction of living structures and research stations (Ziv 2008: 50) without having to transport components to and from Earth.

4.2 New Possibilities

Basalt fibers can, where appropriate, potentially replace both glass and carbon fibers, with advantages in terms of raw material costs and, above all, with regard to the eco-friendliness of the products. "Environmental and healthy working environment concerns lead to basalt fiber being considered as an alternative to glass ones, but significant research and development are still needed […] to make spinning technology more efficient in order to lower the cost of this fiber and make it economically competitive with glass fiber." (Caretto et al. 2017: 8). The use of platinum–rhodium alloy spinners for the production of continuous basalt fibers involves significant production investments both in terms of machinery installation and in terms of operation and maintenance. Ziv (2008: 42) calculates that this component absorbs 65% of the initial set-up cost of the production site. In addition to this investment, it is also necessary to consider the frequent maintenance and/or replacement operations: this type of spinner must be replaced after three or four months, as a reaction occurs between the iron oxides (present in basalt) and platinum that generates important degradation phenomena, including in particular the enlargement of the spinning holes that penalizes the exact dimensional control during production (Caretto et al. 2017: 11). This technological constraint, together with the high cost of the industrial equipment, has, in fact, represented a major brake on any attempts to produce basalt fiber in Italy: despite the presence in this country of numerous and ancient basalt quarries and thanks to a perfectly guaranteed flow of trade on a global level, it has always been preferred to import basalt fiber by exploiting the territorial and productive assets already consolidated mainly in Russia, Ukraine, and China. It is also worth pointing out that, given the considerable mass of the raw material, it is only appropriate to envisage basalt fiber production plants in locations close enough to the material's places of origin, because transporting stones over long distances would not be sustainable.

One of the world's most important basalt deposits is located in western Ukraine. An interesting comparative study (Novitskii and Efremov 2013) found that the properties of basalt fibers obtained from these deposits of andesitic basalt—considered a high-quality raw material—are similar to those of fibers obtained from other types of basalt, in different regions of the world. The study shows that basalt deposits, with SiO_2 content between 40 and 50% (thus suitable for obtaining continuous filaments) are widespread throughout the world because they are used to obtain building stones.

However, rarely have such rocks been used to test the feasibility of obtaining fibers from them. This result is very encouraging because it lays the scientific foundation for future developments of basalt fiber production locally in various parts of the world.

A series of exceptional global events, such as the COVID-19 pandemic and the Russia–Ukraine war scenario, has interrupted or delayed, partially or wholly, trade relations between Europe and these traditional basalt fiber-producing countries. This tragic situation could, however, paradoxically help to rethink the global manufacturing scenario, affecting Italy, thanks to its long tradition in the transformation of mineral raw materials by means of high-temperature furnaces (just think of glassworks, but also cement factories, which are widespread almost everywhere in the peninsula) and its aptitude for innovation in manufacturing (many companies that process high-tech composites are based in Italy). Recent discoveries could in fact make the hypothesis of installing new industrial plants to produce basalt fibers in Italy plausible, with a reduction in initial investment costs: a study carried out for ENEA has shown that traditional platinum–rhodium spinnerets could be replaced with ones made of $MoSi_2$ (molybdenum disilicide), doped with Si_3N_4 (silicon nitride). With this hypothesis, continuous basalt fibers with constant diameter and regular surface morphology were produced, verifying the stability and integrity of the $MoSi_2$–Si_3N_4 spinneret used (Caretto et al. 2017). This is an important scientific fact aimed at reducing the production costs of basalt fibers and encouraging its use with a view to environmental and economic sustainability.

Design, a wide-ranging disciplinary and productive sector of great economic interest for Italy (just think of all the spheres directly or indirectly involved in the so-called Made in Italy, from automotive to furniture, from eyewear to footwear, etc.) can play an important role in the valorization and diffusion of basalt fiber: to do this, however, it is also necessary to carry out research into the esthetic qualities of the material, an important and decisive condition for its acceptance and diffusion at a wider level than the sectors currently involved. Part II of this text is devoted to research topics concerning the artistic and design value of basalt fiber.

References

Agarval BD, Broutman LJ (1990) Analysis and performance of fibre composites, 2nd edn. Wiley, New York, Toronto

Anisoprint (2023). https://anisoprint.com/. Accessed 7 Aug 2023

Arussi E, Reiland J, Ierides M, Borghero L (2023) Composite materials: a hidden opportunity for the circular economy. CSR Europe. https://www.csreurope.org/new-materials-and-circular-economy-accelerator. Accessed 30 Mar 2023

Assocompositi (2022) Circolarità dei materiali compositi. https://assocompositi.it/circolarita-dei-materiali-compositi/. Accessed 7 Apr 2023

Cardone M (2022) Dentro il cantiere della pala eolica riciclabile più grande del mondo. https://economiacircolare.com/pale-eoliche-riciclabili. Accessed 3 Feb 2023

Caretto F (2020) Da rifiuto a risorsa: la nuova vita delle fibre di carbonio. https://www.openin novation.regione.lombardia.it/b/572/ecocarboniocoslafibradicarbonioriciclatadiventaunmater ialeinnovativo. Accessed 10 May 2023

Caretto F, Laera A, Casciaro G (2017) Studio di un materiale ceramico innovativo destinato alla produzione di fibre di basalto. Rapporto tecnico ENEA-RT-2017-22. https://hdl.handle.net/20. 500.12079/6785. Accessed 10 May 2023

Cooke TF (1991) Inorganic fibers. A literature review. J Am Ceram Soc 74(12):2959–2978. https:// doi.org/10.1111/j.1151-2916.1991.tb04289.x

Ecodesign for Sustainable Products Regulation ESPR (2023). https://commission.europa.eu/ene rgy-climate-change-environment/standards-tools-and-labels/products-labelling-rules-and-req uirements/sustainable-products/ecodesign-sustainable-products_en. Accessed 31 Jul 2023

Liu P, Barlow C (2017) Wind turbine blade waste in 2050. Waste Manage 62:229–240. https://doi. org/10.1016/j.wasman.2017.02.007

Novitskii AG, Efremov MV (2013) Technological aspects of the suitability of rocks from different deposits for the production of continuous basalt fiber. Glass Ceram 69:409–412. https://doi.org/ 10.1007/s10717-013-9491-z

Torres JP, Hoto R, Andres J, Garcia-Manrique JA (2013) Manufacture of green-composite sandwich structures with basalt fiber and bioepoxy resin. Adv Mater Sci Eng 2013(Special Issue):Article ID 214506:9. https://doi.org/10.1155/2013/214506

Ziv M (2008) The main problems of basalt fibers producing and using and feasible decision direc-tions. In: BASFA—international workshop on BASalt fiber application, Atti del convegno, Cecina, Polo Tecnologico della Magona, 2007. Collana Ricerca Trasferimento Innovazione. Regione Toscana, Firenze

Part II
Design and Art

Chapter 5
On the Path Toward Applied Research

A sustainable material like basalt fiber has all the prerequisites to be advantageously applied to replace, in some cases, glass and carbon fibers (Fiore et al. 2015). Still, it can have this chance if, in parallel with the study of its properties and the technological development of its applications, there is also a research based on its esthetic acceptance.

5.1 More Ethics

"Nulla ethica sine aesthetica," said the Romans. "Aesthetics is the mother of ethics," Joseph Brodsky said in 1987, during his acceptance speech for the Nobel Prize in Literature:

> On the whole, every new aesthetic reality makes man's ethical reality more precise. For aesthetics is the mother of ethics; The categories of *good* and *bad* are, first and foremost, aesthetic ones, at least etymologically preceding the categories of *good* and *evil*. If in ethics not "all is permitted," it is precisely because not "all is permitted" in aesthetics, because the number of colors in the spectrum is limited. The tender babe who cries and rejects the stranger, or who, on the contrary, reaches out to him, does so instinctively, making an aesthetic choice, not a moral one. (Brodsky 1987)

Yes, but what ethics? And what esthetics? Historically, it is possible to detect a certain fluctuation in the definitions and, especially as far as applied disciplines such as design are concerned, in the interpretations of these two fundamental categories. What are the parameters to be considered when judging a material as ethical, and above all, can these parameters vary over time? As long as an externality is not discovered, i.e., an unintended and normally harmful effect, and as long as there are margins of profit, the world of the industry tends to exploit the wave of technological development and the use of a material (just think of Eternit or the still considerable use of plastics in the toy sector). Here, then, in the realm of esthetics, the judgment

© The Author(s), under exclusive license to Springer Nature Switzerland AG 2023   37
M. Mancini, *The Basalt Fiber—Material Design Art*,
SpringerBriefs in Applied Sciences and Technology,
https://doi.org/10.1007/978-3-031-46102-6_5

of taste—related to the particular historical period—intervenes, shuffling the cards, often putting what is *beautiful* before what is *right*. As early as the late nineteenth century, the sociologist Thorstein Veblen captured exactly the connection between esthetic taste and social expendability.

Andrea Mecacci reports as follows:

> In his 1899 capital work *The Theory of the Leisure Class*, and precisely in chapter VI ("The financial canons of taste"), we can in fact find a decisive illustration of the now definitive links between industrial production and the genesis of social taste. In such a framework, the industrial object is interpreted as a phenomenon whose value lies essentially in its social expendability and, in turn, this social expendability is read as a fundamentally aesthetic problem. Veblen thus thematizes the increasingly strong link between financial wealth and aesthetic taste, defining the object as a means of status exhibition. Few authors have been able to grasp as well as Veblen the overcoming of the use value of the object and the affirmation of an economistic criterion within the object itself [...]. Veblen's basic observation is that bourgeois society in its pursuit of respectability and social honour has transposed these ethical categories also to its own idea of beauty. And beauty, for the bourgeoisie, is first and foremost made visible in objects: it follows that the beauty of an object does not depend on esthetic criteria but on economic evaluations. (Mecacci 2012: 65–66; our translation)

5.2   More Esthetics

This is exactly what happened with carbon fiber: born as a pure high-performance material, used in high-tech contexts, it has become a symbol of exclusivity regardless of the real need to obtain, in the object, absolute performance in terms of lightness and resistance. The carbon-effect texture is so coveted that it is often artificially reproduced in decals or, in the case of more complex shape, through hydrocoating (also called water transfer printing or immersion printing, it consists in by immersing the product in a liquid with a surface film that is then transferred to the object). The aim is to cover other materials in order to give a high-tech image to the product and a consequent esthetic plus. The theme of fascination returns, which acts viscerally on the choice of a product and which, as marketing scholars know, is much more powerful than accurate information on technical, material, and performance characteristics. In many fields, the combination of high-performance demands and carbon fiber is well established: components for the automotive, aeronautical, nautical, aerospace, and sports industries feature carbon fiber structural parts such as car body parts, bicycle frames, and surfboards. In some cases, carbon fiber is chosen more for its esthetic qualities—i.e., social expendability, in the words of the aforementioned Veblen— than for its purely performance-related qualities. Specific, non-structural carbon fiber components are displayed as added value: from motorbike mudguards to standard car rear-view mirrors, these elements alone give a sporty, high-performance value to the product as a whole. In Italy, a country with a solid tradition of craftsmanship and at the same time characterized by design and innovation-oriented industries (typical features of the so-called Made in Italy label), some corporate experiences have been able to exploit the use of carbon fiber in an attractive manner, especially for its esthetic value.

The company Mast Elements (Como, Italy) produces furnishing elements and accessories in carbon fiber, focusing on the quality of the texture, accentuated by light reflections made homogeneous by the resin component of the composite (Mast Elements, 2023). The start-up company Combo Compositi (Compositi Magazine 2023) has patented the Carbit® technology, a coating that makes carbon fiber compatible with food contact, opening up applications in many areas of design and proposing a new material esthetic in areas dominated by ceramics and glass. Ilatro (2023) company (Epi srl, Brindisi) makes suitcases and trolleys in carbon fiber, exhibited and used also and above all for its surface texture value.

Why is basalt fiber not also enhanced esthetically, focusing on its color, textures, light reflections, and relative transparency? Only through a shared social acceptance is it possible to convey messages about the sustainability of material, and this acceptance inevitably passes through the possibility of seeing, touching, and feeling the surface of the material itself. "What is true is also beautiful, regardless of taste, of the preciousness of the object, of formal refinement," as architect Giovanni Michelucci (1997) so agreeably argued: in researching the qualities of the material, the exterior aspect becomes fundamental because it is to all intents and purposes part of the project and its acceptance and also makes its relative ethical values evident through esthetics.

5.3 The Role of Design

Andrea Branzi (2018) carefully reflects on the ethical role of design, predominantly understood as planning endeavor, which is almost always disregarded in practice following the necessary confrontation with the industrial world:

> In the tradition of industrial design there has always existed a strong commitment to dealing with social and environmental problems, with a tendency to collaborate in order to cope with the deformations produced by the savage laws of the market; its very planning methodology, aimed at a drastic semantic reduction, rationalizing forms and functions, was theorized by the English reformist Henry Cole as early as the nineteenth century as the main road to low-cost and aesthetically rigorous industrial products. This ethical tradition placed the design on the frontline of opposition against the excesses of consumerism. [...] This kind of battle led the post-World War II European rationalist movement to try to define (even scientifically) what the real "primary needs" were that industry had to respond to. However, design has never succeeded in this quest for the natural self-regulation of needs, because it has always come up against the unpredictability of the real' needs of contemporary man, material and immaterial needs that do not only have to do with physical necessities but are the result of a non-programmable creative attitude. Moreover, this approach to the issue has conflicted with the operational logic of industry, which has never been rational in nature but has always manifested a tendency to adapt to the laws of the market and competition, producing continuous variants and endless innovative offerings. (Branzi 2018: 57–58; our translation)

Design based on an ecological matrix would thus seem not to exist except at an intentional level. Rather, there is a design that makes products that must satisfy needs, among which are also those of sustainability, ecology, and respect for the environment. It is no longer sufficient to make a product out of a sustainable material if this material is not fit for purpose, is not durable, or does not give the product esthetic dignity: many examples of chairs or tables made of cardboard are detrimental to a correct communication of sustainability, since they are objects destined to have a short life either because of their poor mechanical resistance or their low esthetic value, with the result that this product will become waste to be disposed of quickly (and even the disposal of sustainable, recyclable, or biodegradable material still has a cost in terms of CO_2 and energy). It is therefore not enough to study, research, demonstrate, and verify that basalt fiber is a sustainable material: it is also necessary to investigate how its performance, material, and surface characteristics can contribute to its understanding by users and companies themselves.

5.4   Sensations, Perceptions, and Signs

For proper understanding and thus acceptance, the material should be viewed without hiding anything, for the sake of truth and therefore, to quote Michelucci, of beauty. Since basalt fiber has surface characteristics (chromaticism, reflections, alterations, wefts, weaves) that make it truly interesting, it is worth attempting to enhance them. In other words, it is necessary to attempt an approach, i.e., the inclusion of basalt fiber in a purely semiotic dimension, made up of visible, tangible things, belonging to the field of investigation of perception (De Fusco 2005: 26), an indispensable premise for research between design and art, between mass production and craft production. It is in such spheres of the exhibited, perceptive sphere that the object manifests its meanings and both its first function (denotation)—i.e., the direct effect that a sign produces in the receiver of a message—and its second function (connotation)—i.e., everything that may come to the receiver's mind on the basis of a specific culture. As Barthes (1977: 41) already revealed: "clothes are used for protection and food for nourishment even if they also are used as signs. We propose to call these semiological signs, whose origin is utilitarian and functional, *sign-functions*." That carbon mirror in a standard car has a sign function, that is, its first function is to transmit to the driver the reflection of what is behind the car, and its second function is to arouse the idea of performance, exclusivity, and prestige of the owner. The relationship between function and sign is a non-trivial one that also depends on the characteristics of the material, the sensory ones in the strict sense but also those of association, perceptive, and emotional, as Rossi (2008) describes:

> A product has expressive-sensory characteristics. Like the aspects directly related to the five senses of perception: color, shape, shine, softness, etc., there are also association characteristics, i.e. the idea that is associated with the product (e.g. we associate the idea of luxury with gold) and perceptual characteristics, which are the reactions that a person has to a product or a material (modern, fashionable, sophisticated, fun, easy to use, etc.). Finally, we have the emotional characteristics, which concern the feelings that a product or material provokes, at first, to see it and desire it and, finally, to possess it after buying it: happiness, sadness,

and pride. Considering these aspects, it is very clear that when designing a new product, the choice of materials to be used must not only consider mechanical properties, resistance to wear and corrosion but must also take into account other aspects. (Rossi 2008: 23; our translation)

5.5 Research Themes

The range of applications for basalt fiber confirms that this high-performance and sustainable material is always used in a concealed manner: inside panels, as a structural layer to be covered, as a reinforcement to be wrapped with other products, as a layer in layered elements, etc.

There are very few cases in which this material is granted the dignity of being visible. These include the research work of the designers Anja Zachhau and Jakob Kukula (Fig. 5.1a, b), who have used visible basalt fiber as an outer layer in flame-resistant protective clothing, as a substitute for highly carcinogenic asbestos fiber, formerly used (Hafsa and Rajesh 2016), or the US company Myrdal Orthopedics Technologies (2023), which also uses visible fiber for its orthopedic components, or the company Sanded Australia (2023), which in part uses the visible material for surfboards and accessories.

In the period 2020–2022, I proposed to my students of the Design Culture course—master's degree in New Expressive Languages at the Accademia di Belle Arti (Fine Arts Academy) in Florence, Italy—to tackle the challenging topic of a research on basalt fiber (Fig. 5.2). All the involved artists/designers worked on the same starting material, consisting of a piece of basalt fiber in a flat-weave fabric (taffetas). In this kind of fabric, the warp and weft alternate top and bottom and vice versa (Frassine et al. 2008: 111). The aim was to analyze, interpret, and enhance the potential of the basalt fiber *skin* in search for new margins of application in areas more suited to the work of designers, artists, and craftsmen. In fact, the links between research in the field of art and that of design are very close, and the reciprocal boundaries are anything but precise and delineated. The experiences realized are narrated, thanks also to Juri Ciani's photographs, within four macro-themes: texture, volume, light, and proposals. In spite of the various differences in theme, approach, development, and design outcomes, this research shows that the margins for the use of this material are still very wide and include diverse sectors: usage products, furnishings, lighting fixtures, theatrical objects, wall coverings, symbolic objects, etc.

a

b

◄**Fig. 5.1 a** Made of Rock, by Anja Zachau, Jakob Kukula. Courtesy of Anja Zachau. **b** Made of Rock, by Anja Zachau, Jakob Kukula. Courtesy of Anja Zachau. The authors of the "Made of Rock" project (Bauhaus University, 2017) focused on the extreme heat resistance of the basalt fiber, studying two concepts of protective clothing, a heat-resistant protective glove and a fireproof jacket, thanks also to the research on the material developed at the Saxon Textile Research Institute in Chemnitz. Both products consist of three layers: basalt fiber, insulation, and a lining fabric for maximum user comfort

Fig. 5.2 Concepts, prototypes, and material tests during classroom work with basalt fiber at the Florence Academy of Fine Arts, 2022. Picture by the author

References

Barthes R (1977) Elements of semiology (trans: Lavers A, Smith C). Hill & Wang, New York

Branzi A (ed) (2018) Il design. Storia e controstoria. Giunti, Firenze-Milano

Brodsky J (1987) Nobel lecture, 8 December 1987. https://www.nobelprize.org/prizes/literature/1987/brodsky/lecture. Accessed 9 May 2023

Compositi Magazine (2023). https://www.compositimagazine.it/combo-s-r-l-crea-carbit-la-fibra-di-carbonio-e-i-materiali-compositi-incontrano-lalimentare/. Accessed 7 Aug 2023

De Fusco R (2005) Una semiotica per il design. Franco Angeli, Milano

Fiore V, Scalici T, Di Bella G, Valenza A (2015) A review on basalt fibre and its composites. Compos B Eng 74(1):74–94. https://doi.org/10.1016/j.compositesb.2014.12.034

Frassine R, Soldati MG, Rubertelli M (2008) Textile design. Materiali e tecnologie. Franco Angeli, Milano

Hafsa J, Rajesh M (2016) A green material from rock: basalt fiber—a review. J Text Inst 107:923–937. https://doi.org/10.1080/00405000.2015.1071940

Ilatro (2023). https://www.ilatro.com/IT/. Accessed 7 Aug 2023

Mast Elements (2023). https://mastelements.com/. Accessed 7 Aug 2023
Mecacci A (2012) Estetica e design. Il Mulino, Bologna
Michelucci G (1997) Dove si incontrano gli angeli. Fondazione Michelucci, Carlo Zella editore, Fiesole (FI)
Myrdal Orthopedics Technologyies (2023). https://www.myrdalorthopedics.com/. Accessed 7 Aug 2023
Rossi S (2008) I rivestimenti. La pelle del design. Alinea, Firenze
Sanded Australia (2023). https://www.sanded.com.au/ Accessed 7 Aug 2023

Chapter 6
The Texture

The basalt fiber produced in continuous filament form is also a textile material that can be subjected to several mechanical stresses both during production and application (Bonetti et al. 2012: 87). It is important to remember that fibers are defined as solid materials in an elongated, thin, and flexible form, but this does not automatically mean that all fibrous materials are also textiles (Chawla 1998). The dimensional characteristics (length, thickness, density, cross-section) and external characteristics (brightness, hand-feel, softness, bulkiness) of the product have an important influence, so much so that even a slight modification of one or more parameters can clearly transform the perception. In fact, "materials play a dual role: they support the technical functionality and simultaneously create the personality of the product" (Ashby and Johnson 2002: 2, our translation). The research work presented in the following pages focuses on terms such as interpretation, addition, subtraction, weave, embroidery, color, and filter.

The role of the material's reflections is particularly important: the uniformity of the surface of each fiber generates a high brightness, and the reflections generated by the fiber in its natural state (without resins) can produce fascinating luminous effects, in a dark brown/gold color base that is very reminiscent of certain metals. The research into the material also led to experimenting with the possibility of coloring the fibers (Bona et al. 1981) with different types of pigments and, given the material's regularity and resistance, testing its use as a mask or stencil.

6.1 The Fiber Molting

The exploration of basalt fiber weaving guides Elisa Pietracito's research path, as she says: "as I moved the fiber, the effect of light that was created made me think of scales. I started experimenting by trying to embroider the fabric with threads from the fiber itself, but the basic weave opened up, forming spaces. So I decided to proceed by subtraction, reflecting on the ethical aspect of not adding to the existing. The subtraction of the filaments takes place at different rhythms, following organic spot shapes, replicating the design of the open skin of snakes, naturally suggested by the reflective property of the material."

The artist meticulously analyzes filament reflection and fiber texture at various illumination modes (Figs. 6.1a–c), then experiments with different types of subtraction of filaments: alternating in vertical (Figs. 6.2a–c), in vertical and horizontal (Fig. 6.3), designing waves (Fig. 6.4) and curves (Fig. 6.5), and designing texture from rhythmic (Fig. 6.6) or specular subtraction of filaments, going so far as to test the possibility of drawing geometric shapes (Fig. 6.7) and the technical feasibility of basic forms of embroidery (Figs. 6.8a–c, and 6.9). It is a minimal work that exploits the organicity of the weave-color-embroidery system, enhancing the value of the fiber in an absolute sense, in its natural state, and without resins. The point of view is that of an investigation with the intent of ideation and execution of a fabric collection with reference to product identity (Pompas 2003).

Elisa Pietracito. Borgo San Lorenzo, 1998.

Her research is characterized by respect and attention to contemporary issues related to landscape and sustainability in art, combining a fascination for the nature of materials of organic or recycled origin with refined manual workmanship. She has collaborated with Murate Art District, Museo della Ceramica di Montelupo, Fondazione Palazzo Strozzi, Estuario Project Space, TIMEOUTvibes, FAF Female Artists in Florence, SINCRESIS arte, Chille della Balanza, Festival Foglia Tonda, NAM Not a museum, Manifattura tabacchi, and Effetto Venezia.

6.2   Decoration Tests

Jasmine Morandini's research focuses on the technical and expressive potential of using color directly on the fiber, without resins. The artist tells: "my project consists of decorative tests carried out on basalt fiber to test how colors can relate to the aesthetics of this material. With a spatula, I painted marks or color layers giving importance to the materiality of the technique used: blue oil color, acrylic—red, black (Fig. 6.10), and yellow (Fig. 6.11)—and white plaster white. The fiber's texture creates peculiar effects of light and shadow, reminiscent of the characteristics of metals. Another test I carried out consisted in using the fiber as a stencil: by placing it on a sheet of paper, I spread a blue acrylic layer (Fig. 6.12), resulting in a dotted pattern (Fig. 6.13a, b) reminiscent of the Pop Art period."

From the compatibility of basalt fiber with many types of substances, already exploited for technical applications (with all kinds of resins, plasters, and concretes) comes the potential to interface with pigments of various kinds, opening up the field to interesting developments and margins for future research in many contexts where the chromatic element and the visual effect must also meet stringent requirements in terms of fire resistance, such as the fields of stage design, installations, and exhibition design.

Fig. 6.1 **a** Basalt fiber fabric with perpendicular light source. **b** Basalt fiber fabric with top light source. **c** Basalt fiber fabric with scattered light. Pictures by Juri Ciani

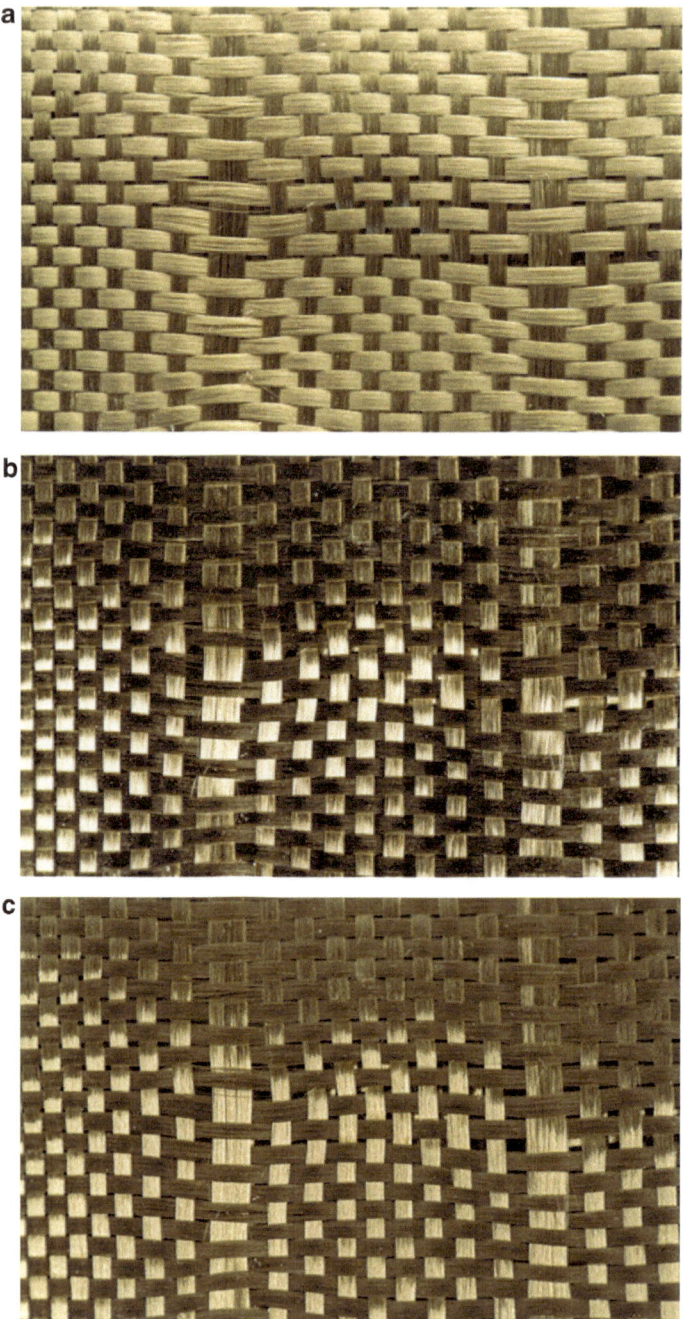

Fig. 6.2 **a** Basalt fiber fabric with alternate vertical filament subtraction, in scattered light. **b** Basalt fiber fabric with alternate vertical filament subtraction, with bottom light source. **c** Basalt fiber fabric with alternate vertical filament subtraction, with scattered/bottom light source. Author: Elisa Pietracito, pictures by Juri Ciani

Fig. 6.3 Basalt fiber fabric with alternate vertical and horizontal filament subtraction, forming waves in the bottom. Author: Elisa Pietracito, picture by Juri Ciani

Fig. 6.4 Basalt fiber fabric with double-vertical-filament subtraction, forming waves. Detail with macrolens. Author: Elisa Pietracito, picture by Juri Ciani

Fig. 6.5 Basalt fiber fabric filaments forming waves and curves, overlaid with such a fabric. Author: Elisa Pietracito, picture by Juri Ciani

Jasmine Morandini. Pisa, 1998.

Her research focuses on representing and denouncing the visible consequences of climate change on the landscape, using mainly natural and recycled materials. She took part in "Perenne Attualità," Palazzo Strozzi (Florence), curated by the Istituto Marangoni Firenze; in "L'inconsolabile" by Francesco Carone, Villa Pacchiani (Pisa), curated by Ilaria Mariotti; in "Mind in the Map," Not a Museum (Florence), curated by Robert Pettena, Stefano Giuri, and Gabriele Tosi; and in "Il Gioco della Natura" curated by Gaia Bindi and installation by Alessandro Scilipoti.

6.3   Tapestry

Letizia Lo Verde's work is based on research into the material expressiveness of fiber texture through a study of the pattern effects generated by its use as a stencil. In this research, the fiber's characteristic of resistance to abrasion and compatibility with pigments and substances is exploited to create a reusable matrix, thinking of a limited series production. From the relationship between texture and color come new designs that can completely change the appearance of a fabric. There is thus the possibility of "drawing with color," which allows us, even by weaving a simple canvas, to decorate it with a wide range of patterns (Nieuwenhuijs 1997).

The artist says "I cut a strip of basalt fiber and widened the mesh of the material to create openings. I then placed the strip on a sheet, using it as a template to create

Fig. 6.6 Basalt fiber fabric texture composition obtained from vertical alternating rhythmic subtraction of filaments (1-2-1-2-1-3-1-5-1-5-1-1). Author: Elisa Pietracito, picture by Juri Ciani

an additional color test board; I mixed acrylic color with vinyl glue and then spread it over the strip with a sponge (Figs. 6.14 and 6.15). The previously created openings allowed the passage of color over the sheet and therefore the creation of new images." The mix of color and glue creates very interesting textural and chromatic alterations on the basalt fiber (Figs. 6.16a–c). "Finally, I placed the strip—used as a template—on a different board, which retained parts of the color used. The texture created (Figs. 6.17a, b, 6.18a, b, 6.19), inspired by references to oneiric imagery, has many possibilities for application (tiles, fabrics, tapestries, wallpaper or other)."

Letizia Lo Verde. Florence, 1998.

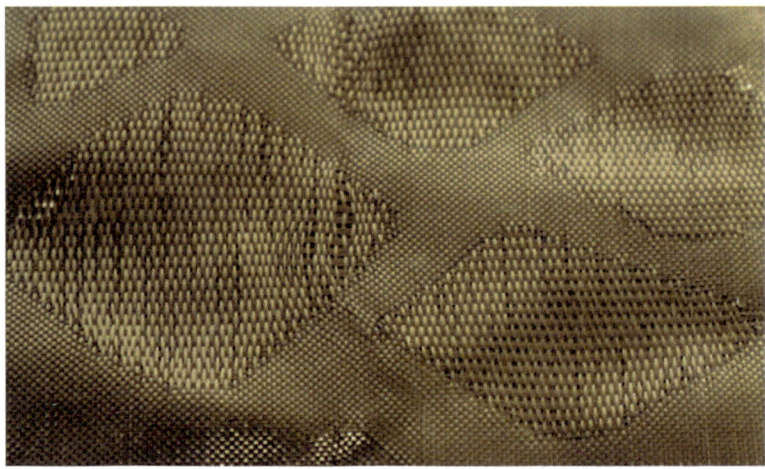

Fig. 6.7 Basalt fiber fabric subtraction finalized at obtaining geometrical figure on the texture. Author: Elisa Pietracito, picture by Juri Ciani

The development of her artistic path is guided by pure and simple experimentation with various types of materials in which the forms created arise freely and spontaneously, without foreseeing any possibility of reproduction. The inspiration is that of the everyday context, changing in relation to sounds, colors, and images as well as in relation to different local cultures. Sensitivity thus becomes the founding value of each of her artistic projects, born from her curiosity to discover new materials and techniques, with the help of the photographic tool. The artist collaborates with artistic workshops aimed at inclusion such as "Sharing Europe" and "Scienza in Fabula."

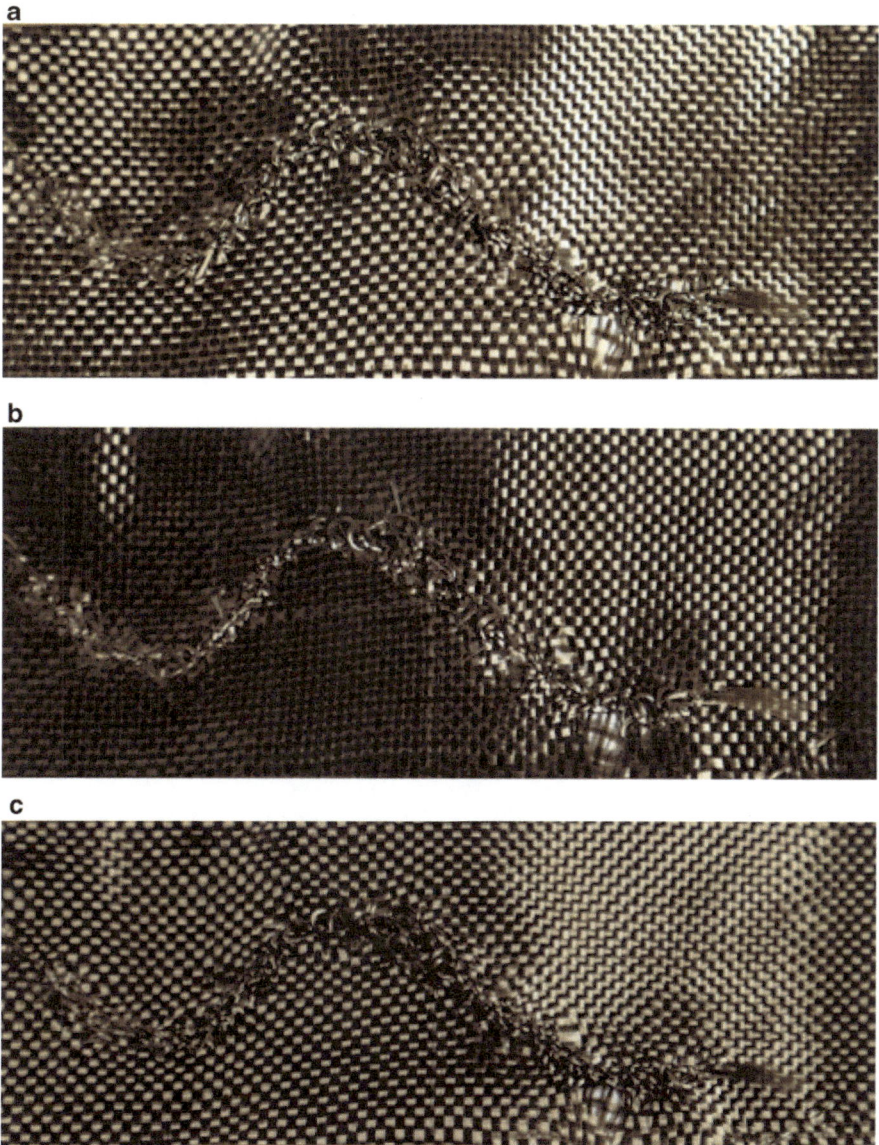

Fig. 6.8 a Basalt fiber fabric with embroidered design, photographed with perpendicular light. **b** Basalt fiber fabric with embroidered design, photographed with light source from the right. **c** Basalt fiber fabric with embroidered design, photographed with scattered light. Author: Elisa Pietracito, pictures by Juri Ciani

Fig. 6.9 Embroidered design on basalt fiber fabrics: detail with macrolens. Author: Elisa Pietracito, picture by Juri Ciani

Fig. 6.10 Black acrylic color painted over basalt fiber fabrics. Courtesy by Jasmine Morandini

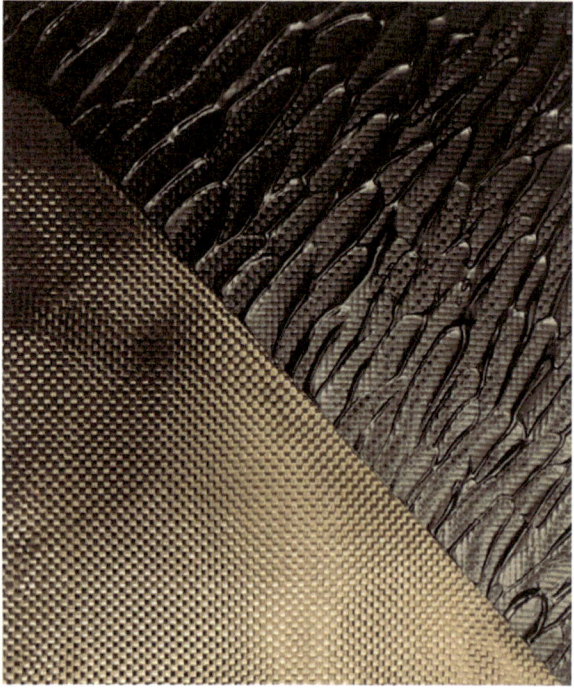

Fig. 6.11 Yellow acrylic color painted over basalt fiber fabrics. Courtesy by Jasmine Morandini

Fig. 6.12 Blue acrylic layer over basalt fiber fabrics. Author: Jasmine Morandini, picture by Juri Ciani

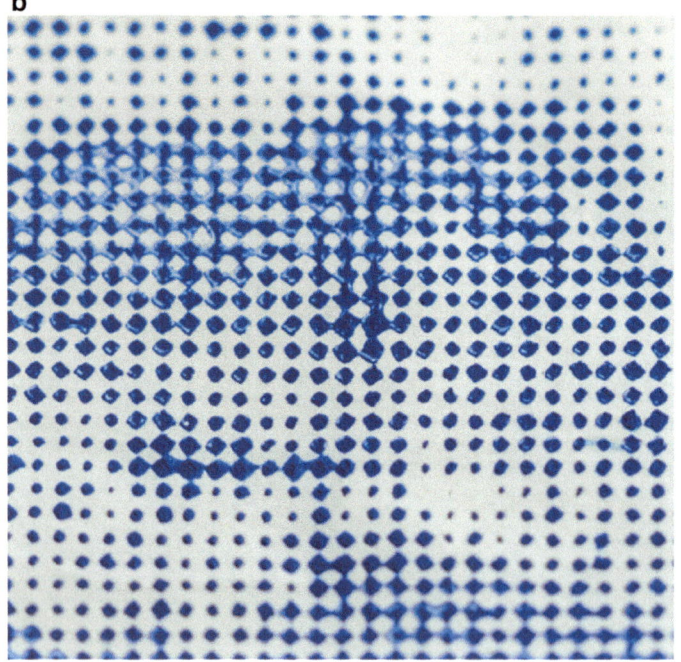

Fig. 6.13 a Dotted pattern in blue acrylic color, generated from the usage of basalt fiber fabrics as a stencil. **b** Detail with macrolens. Author: Jasmine Morandini, pictures by Juri Ciani

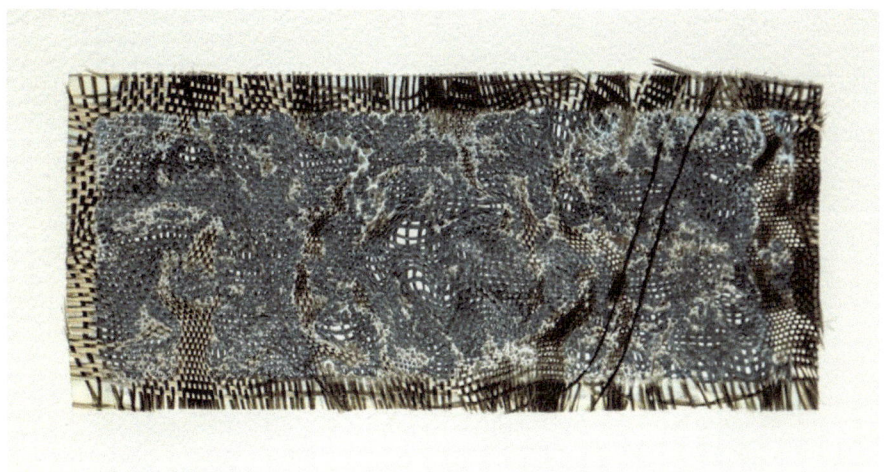

Fig. 6.14 Acrylic color with vinyl glue spread over a strip of basalt fiber fabrics. Author: Jasmine Morandini, pictures by Juri Ciani

Fig. 6.15 Detail with macrolens

Fig. 6.16 **a** Acrylic color with vinyl glue spread over a strip of basalt fiber fabrics. Detail 1 **b** Detail 2 **c** Detail 3. Author: Jasmine Morandini, pictures by Juri Ciani

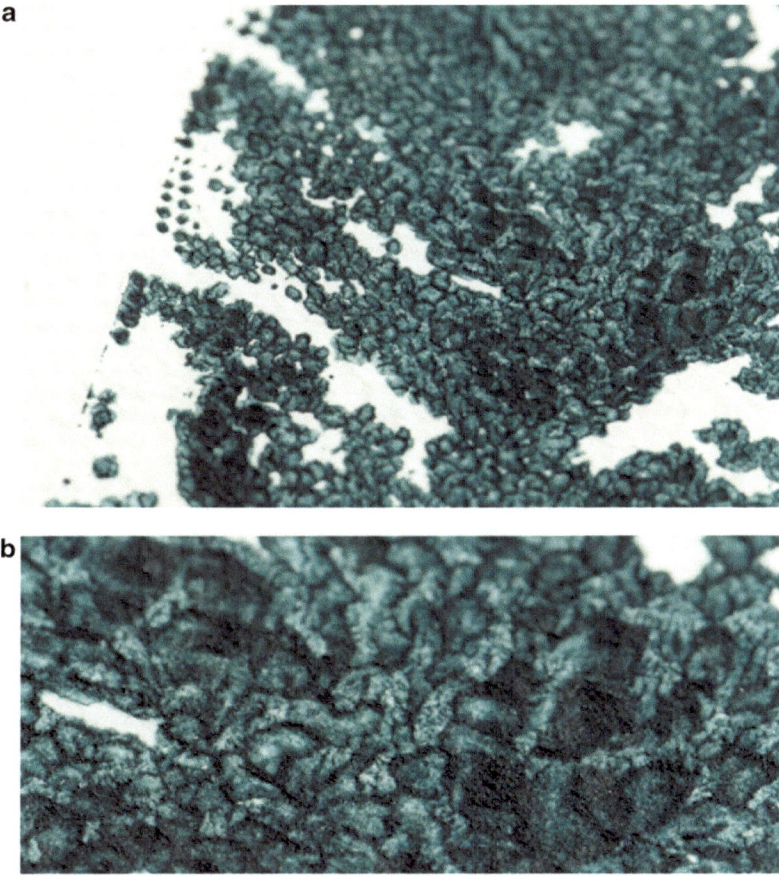

Fig. 6.17 **a** Texture generated by the passage of acrylic color with vinyl glue through a strip of basalt fiber fabrics (type 1). **b** Detail with macrolens. Author: Jasmine Morandini, pictures by Juri Ciani

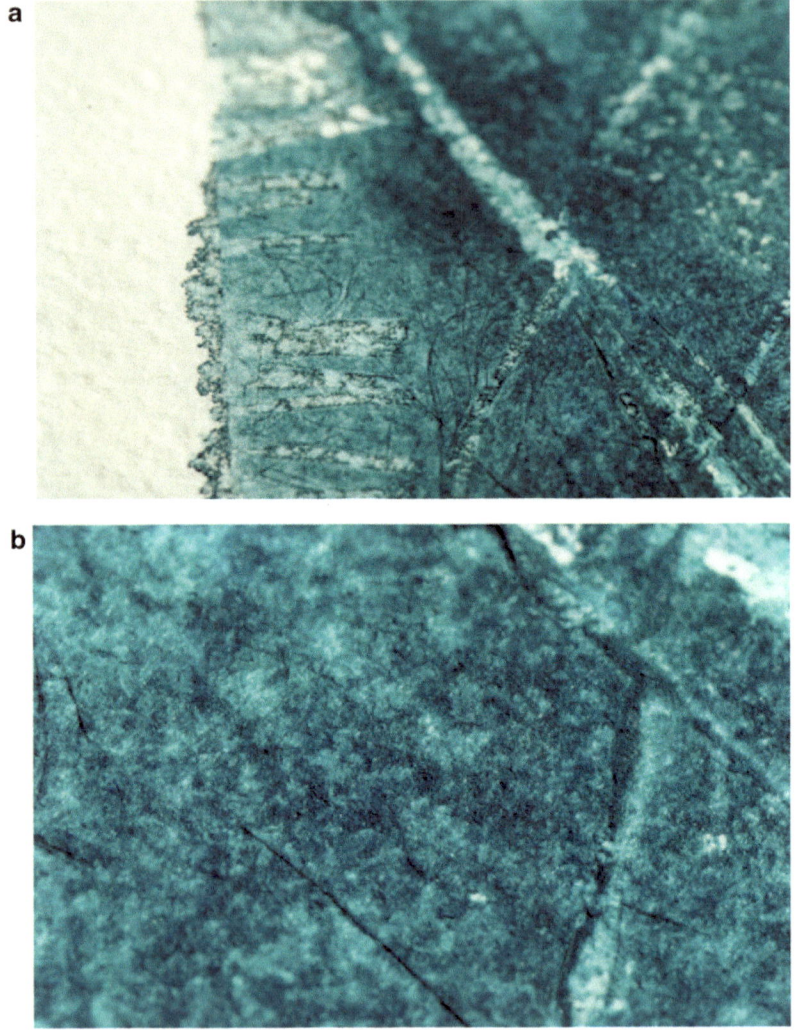

Fig. 6.18 a Texture generated by the passage of acrylic color with vinyl glue through a strip of basalt fiber fabrics (type 2). **b** Detail with macrolens. Author: Jasmine Morandini, pictures by Juri Ciani

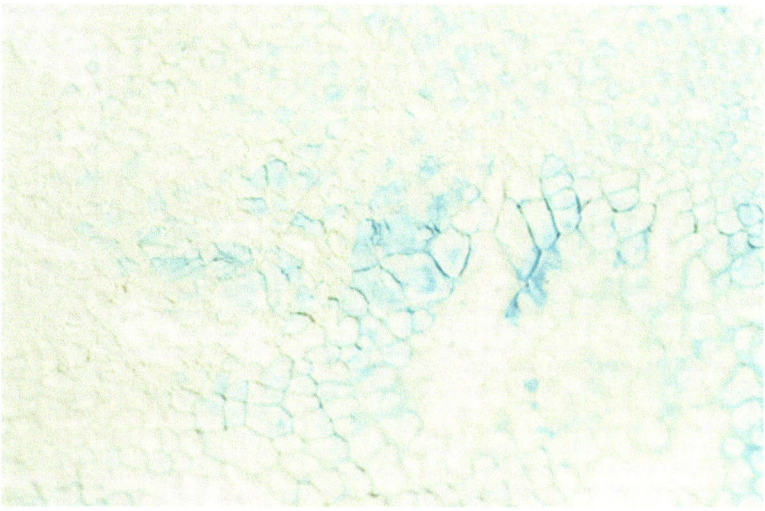

Fig. 6.19 Texture generated by the passage of acrylic color with vinyl glue through a strip of basalt fiber fabrics (type 3). Author: Jasmine Morandini, picture by Juri Ciani

References

Ashby MF, Johnson K (2002) Materials and design, Elsevier ltd, Kidglinton, England. Italian edition: Ashby MF, Johnson K (2005) Materiali e design. L'arte e la scienza della selezione dei materiali per il progetto (trans: Del Curto B, Levi M, Pedeferri MP, Rognoli V), Casa editrice ambrosiana

Bona M, Isnardi FA, Straneo SL (1981) Manuale di tecnologia tessile. Zanichelli, Bologna

Bonetti F, Dotti S, Tironi G (2012) Fibre tessili. Struttura, caratteristiche, proprietà. Tecniche Nuove, Milano

Chawla KK (1998) Fibrous materials. Cambridge University Press

Nieuwenhuijs M (1997) L'intreccio ed il colore nei tessuti a due e quattro licci. Il Castello, Milano

Pompas R (2003) Textile design. Hoepli, Milano

Chapter 7
The Volume

The aim of this line of research on basalt fiber was to verify the possibility of generating volumetric shapes using various techniques: papercraft-type processing, molds obtained from existing products, molds made specifically for the project, various molding techniques, by exploiting pressure or gravity. During the research, it was possible to verify both the limitations of the material and its potential. Limitations, for example, include the difficulty of making sharp bends because of the risk of breaking the fibers; among the potentialities, it emerged the possibility of playing with the relative translucency to obtain light effects, as detailed in the next chapter. The experiments on volume also made it possible to understand the nature of the fabric, which is not elastic but nevertheless resistant to rumpling; resistance to rumpling is the ability of a fiber to resist the formation of creases with the recovery or not of the initial shape after the deformations it has undergone (Bonetti et al. 2012: 109). The chosen basalt fiber textile fabrics can adapt to follow volumetric or surface changes, even alternating between solids and voids. In terms of the hand the fabric used is rather supportive, smooth but not yielding; the term *hand* or *hand feel* refers to a set of characteristics such as softness, suppleness, and luminosity. A hand that is firm to the touch is stiff and not very soft, a soft hand is tender and elastic to pressure (Bonetti et al. 2012: 93). To the touch, while somewhat reminiscent of certain metallic textures, it differs from metal in that it is not cold: as it is not conductive, it remains inert without subtracting heat from the body.

7.1 Free Forms

"The ultimate object of design is form" (Alexander 1964: 15). Starting from this consideration, Marco Mancini's research work focuses on experimenting with different possibilities of volumetrically shaping the starting material, in combination with epoxy resin (Akovali 2001), in order to verify the different texture qualities, based on the shapes that can be obtained. The research path was carried out keeping in mind the logic that "the most direct connection between material and form is given by the stresses that materials can withstand (Ashby and Johnson 2002: 99)."

In the image on Fig. 7.1a, the solids are obtained from a sheet of basalt fiber fabrics laminated with epoxy resin and then cut and folded using simple papercrafts techniques: this made it possible to verify the maximum bending angles obtainable without damaging the fiber as well as the possibilities of cutting and grafting surfaces together. The texture of the fiber-resin composite was treated in different ways: in Fig. 7.1b it was treated with polish-type cream, in Fig. 7.1c with sandpaper and in Fig. 7.1d left untreated.

In the image on Fig. 7.2a the volume was modeled from a mold, which was previously treated with a release agent. Once released, the product was then repeatedly compressed and crumpled: these operations have no effect on the volume, which always returns to its initial state thanks to the resilience of the fiber, while they do have an effect on the resin component, which detaches, generating veins reminiscent (Fig. 7.2b) of those of certain stone materials.

The product on Fig. 7.3 was modeled by placing the freshly resin-impregnated sheet over an air-inflated balloon which, once the resin had dried, was punctured, quickly releasing the shape. In addition, shredded fragments of recycled rubber were inserted before the resin dried to give greater rigidity to the whole as well as a different color appearance.

One of the aims of this line of research has also been to test new possibilities in established composite processing technologies (Anderson and Tushman 1990). In the concept on Fig. 7.4a, b the experimentation concerned the technique used for lamination, which was carried out by diluting the epoxy resin with acetone and spraying it with an airbrush: this technique, scarcely mentioned in technical literature, has been borrowed by boat repairers, who use it for small, localized maintenance tasks. In fact, lamination is normally carried out by spreading the sheet of fiber on a flat surface, impregnating it with resin on both sides and then forming it on the mold; since the resin starts to harden very quickly, the time to adapt the artifact to the mold is short and leaves no margin for error. By means of the spray technique, on the other hand, it is possible to spread the sheet of fiber even in complex shapes with due calm and precision, and only then to apply the resin with the added advantage of being able to apply it only on one side, thus saving material, making it lighter and allowing the possibility of obtaining finishes with different aspects.

In the object on Fig. 7.5a, b the research focused on the production of a thin, self-supporting form, thanks to ribs obtained from folds formed before the resin dried. Even in this case, no mold was used but a simple air-inflated balloon.

In this circumstance, the intent was to reproduce the sinuous and flowing forms that characterize basalt lava. In following the fascination that motivated the research path underlying this book, a reference to the ancestral link between the four elements of classical tradition (Air, Water, Earth, Fire) responsible according to Empedocles for the union and separation of things (O'Brien 2007) was attempted. In this case, air gives rise to the form and the earth (rock) provides the material, shaped through fire.

Fig. 7.1 **a** Solids obtained from a sheet of basalt fiber fabrics laminated with epoxy resin and then cut and folded using simple papercrafts techniques. **b** Detail of surface treated with polish-type cream. **c** Detail of surface treated with sandpaper. **d** Detail of surface of the composite material left untreated. Author: Marco Mancini, pictures by Juri Ciani

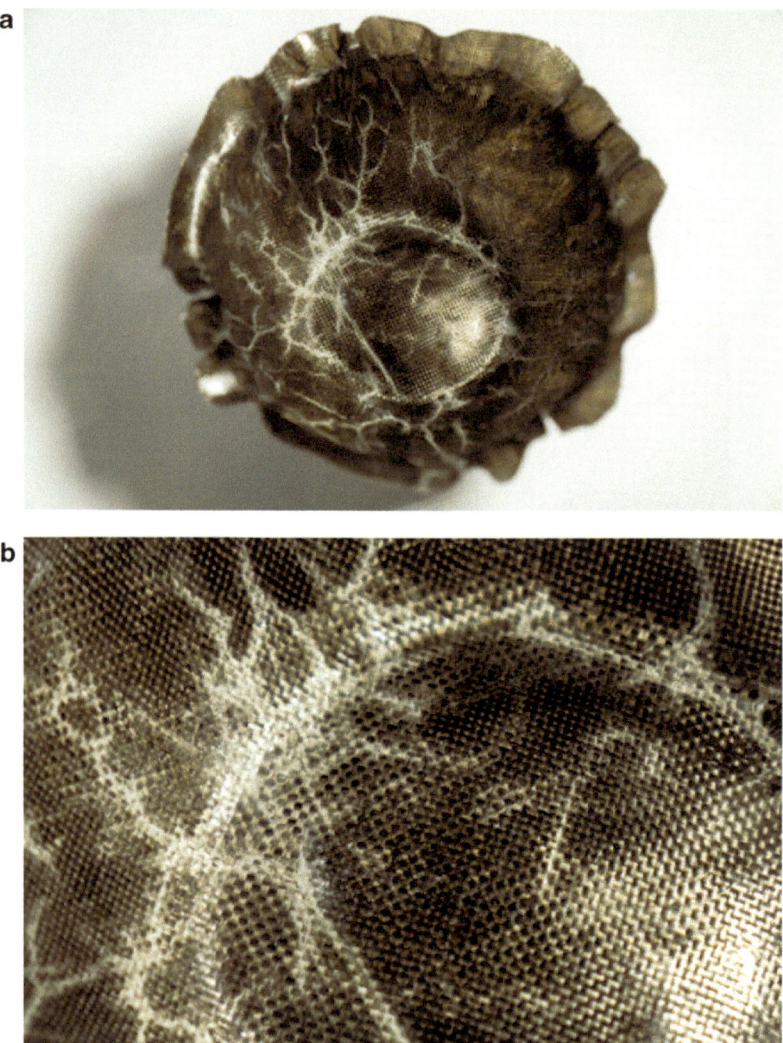

Fig. 7.2 a Volume modeled from a mold, previously treated with a release agent. Once released, the product was then repeatedly compressed and crumpled. **b** Details of veins generated from the detachment of the resin. Author: Marco Mancini, pictures by Juri Ciani

7.2   Empty Versus Full

The research work is based on the possibility of exploiting basalt fiber for its sound-absorbing characteristics (Dlugosz et al. 2015) and adaptation to molds of complex shapes. Benedetta Chiari's material experimentation is therefore aimed at obtaining a remarkably three-dimensional product, alternating concave and convex surface

Fig. 7.3 Object modeled by placing the freshly resin-impregnated sheet—with addiction of shredded fragments of recycled rubber—over an air-inflated balloon which, once the resin had dried, was punctured. Author: Marco Mancini, picture by Juri Ciani

portions (Fig. 7.6), which can be repeated and manageable in many variations. As an alternative to traditional smooth panels, this experimental module, light and resistant in form, acts as a self-supporting shell that can be installed on many types of existing curtain walls, it can be filled with basalt wool or other material (depending on the performance required) and is suitable both for visible use and, as verified in the texture research experiments, for treatment with pigments. This experimentation makes it possible to think about the proposal of laminates both at a craft and artistic level and at an industrial level, varying the dimensions of the panel and the number and characteristics of the solid/void ratio.

Benedetta Chiari. Fucecchio (Florence), 1998.

She participates in workshops and artist residencies, establishing a collaboration with her colleague Elisa Pietracito. Her works are presented as assemblages of poor materials, often taken from the natural context and reworked also in a graphic key. She has collaborated with various realities such as Carico Massimo, Chille de la Balanza, Murate Art District, and Villa Pacchiani.

7.3 In the Wind

In the case of artistic events opened to the public, such as theater performances, scenic demands must coexist with even rigorous regulatory requirements. This gave rise to the idea of using the natural heat and fire resistance of basalt fiber to propose

Fig. 7.4 **a** Lamination was
carried out by diluting the
epoxy resin with acetone and
spraying it with an airbrush.
b Detail of the surface.
Author: Marco Mancini,
pictures by Juri Ciani

Fig. 7.5 a Self-supporting
form, thanks to ribs obtained
from folds formed before the
resin dried. Instead of a
mold, an air-inflated balloon
was used. **b** Detail. Author:
Marco Mancini, pictures by
Juri Ciani

Fig. 7.6 Experiment with basalt fiber fabrics modeled alternating concave–convex surface portions.
Courtesy by Benedetta Chiari

fireproof objects, furnishings, and accessories for the stage, with the advantage of
also exploiting the natural shine of the material to generate lighting effects that can
be managed with the lighting systems used in theaters or performance halls. The
headgear study (Fig. 7.7a–c) presented here is made by shaping and weaving the
sheet of basalt fiber onto a form, then securing the free ends of the threads with vinyl
glue. Vinyl glue is much easier and quicker to handle as well as easier to store than
the resins traditionally used in industrial lamination. It is particularly suitable where
a high final strength of the product is not required and when repairs or modifications
have to be carried out quickly, even in indoor or poorly ventilated environments, due
to its absence of unpleasant odors. The verification of compatibility between basalt
fiber and vinyl glue opens up the possibility of use in many creative and artistic fields,
such as workshop-type activities, even for children.

Luisa Nacci. San Miniato (Pisa), 1998.

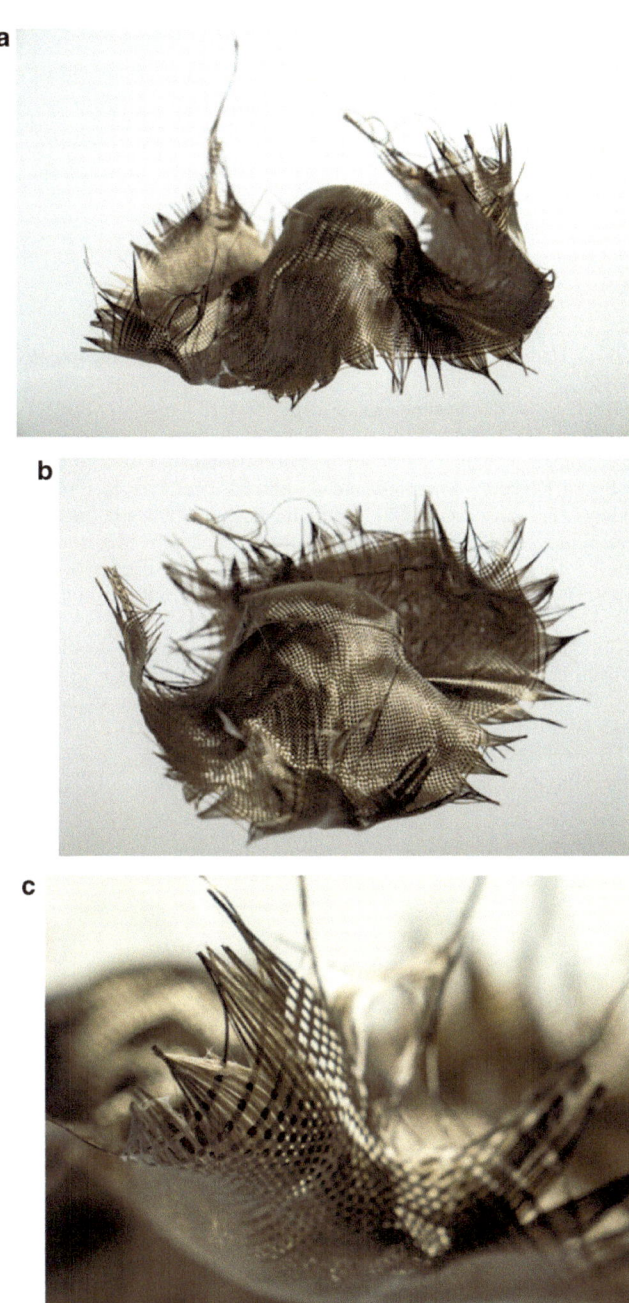

Fig. 7.7 a Headgear study, made by shaping and weaving the sheet of basalt fiber fabrics onto a form, then securing the free ends of the threads with vinyl glue. **b** Headgear study, different view. **c** Headgear study. Detail. Author: Luisa Nacci, pictures by Juri Ciani

Her creative path, based on color and materiality, has developed mainly through video art and performance, with strong influences from the world of classical dance. She has taken part in events and exhibitions linked to her home territory, including "Immersed in nature," the personal exhibition "Fra-Mentazione" for the Meyer Hospital in Florence, "We are seeds…," "Transparency" at Essenza (Siena).

References

Akovali G (2001) Handbook of composite fabrication. RAPRA Technology Ltd, Shawbury, Shropshire

Alexander C (1964) Notes on the synthesis of form. Harvard University Press, Cambridge (MA)

Anderson P, Tushman ML (1990) Technological discontinuities and dominant designs: a cyclical model of technological change. Adm Sci Q 35(4):604–633

Ashby MF, Johnson K (2002) Materials and design. Elsevier Ltd, Kidglinton, England. Italian edition: Ashby MF, Johnson K (2005) Materiali e design. L'arte e la scienza della selezione dei materiali per il progetto (trans: Del Curto B, Levi M, Pedeferri MP, Rognoli V), Casa editrice ambrosiana

Bonetti F, Dotti S, Tironi G (2012) Fibre tessili. Struttura, caratteristiche, proprietà. Tecniche Nuove, Milano

Dlugosz W, Kirchgessner G, Golovatchev J (2015) Innovation and development of the woven and non-woven basalt fiber products. In: Basalt fiber: challenges and development potential. Nuremberg, December 2015. https://www.researchgate.net/publication/299595091_INNOVA TION_AND_DEVELOPMENT_OF_THE_WOVEN_NON-WOVEN_BASALT_FIBER_ PRODUCTS. Accessed 10 Aug 2023

O'Brien D (2007) Empedocle, in Il sapere greco. Dizionario critico, vol 2. Einaudi, Torino

Chapter 8
The Light

Light has always been a field of experimentation for designers and artists, as well as a special area of mutual contamination between art and design, between technical requirements and expressive values. There are many cases of designers who became famous through the design of a lamp, just as some lamps are often considered objects of artistic conception (Castelli et al. 2007). The contamination between design and art often generates objects that take on a value that, for the user, goes beyond the logic of pure function. Although we are used to coexisting with the products we use, using them properly, and relating to them immediately even in the case of new objects (Turri 2011), when a product comes to take on a deeper meaning, because it involves the emotional and perceptual side, then the durability of the product also becomes an important parameter; this is one of the most important chances to combat the various forms of obsolescence (programmed, planned, symbolic) that are unfortunately adopted by manufacturers to instill in the buyer the need of a new product (Latouche 2012).

Whether it is a question of searching for new reflections and shimmers within the texture or using the material as a diffuser/medium between the light source and the environment, basalt fiber seems to lend itself well to being used in this context. Due to the extreme homogeneity of its fibers, the basalt fabric has its own natural brightness, and its reflections are shimmering both according to the type of illumination and the location of the material. Its heat-resistant characteristics also make it a material suitable for close contact with light sources without the danger of flammability. During the research work it became apparent that, although the fiber is opaque, light waves can pass through the textile weave: this produces a relatively translucent effect, making the material suitable for use as a light diffuser on which drawings or graphics can be realized. Different types of surface treatment can contribute to a filter effect for certain wavelengths. The works proposed in this chapter include models made of natural material, composite material with epoxy resin, and in combination with resin and pigments.

© The Author(s), under exclusive license to Springer Nature Switzerland AG 2023   73
M. Mancini, *The Basalt Fiber—Material Design Art*,
SpringerBriefs in Applied Sciences and Technology,
https://doi.org/10.1007/978-3-031-46102-6_8

8.1 Vase Table Lamp

Through the research of new expressive possibilities offered by the material, the proposed model combines the typological characteristics of a table object with the more functional ones of an illuminating accessory (Fig. 8.1a–c). Through light effects, the hybrid object obtained emphasizes the chromatism of the fiber, made brighter by the combination with resin. Technically, the product was made by wrapping the sheet of fiber around an internal structure, which was then removed, and distributing the resin with a brush, thus obtaining a composite material (Lee 1991); the LED strip also has the function of closing the envelope.

Luo Shihua. China, 1995.

Member of the China Kunming Folk Artists Association, she published articles for the magazine "Appreciation," based in Shan Xi Province. In 2022, she exhibited his work *Marcire* at "Mind the Map" in Florence.

8.2 Sunbeam

The proposed model is a wall lamp with the function of a companion light for the night (Fig. 8.2a–c). The shape is reminiscent of a cloud through which the sun's rays can filter, attenuated. For this reason, a lighting system with a manual light temperature adjustment system was chosen to recreate the shades of natural light according to the weather conditions. Formal and material research has generated an object made of different materials: the basalt fiber is wrapped around an essential wire frame, positioned knife-edge to the wall, with the conceptual and technical intention of capturing, delimiting, and enclosing the light, allowing it to escape in a direct but controlled manner, grazing the wall. To complete the surface exposed to the user, there is a weave of another natural fiber, wool, symbolically, and perceptually similar to the softness of the cloud, through which the light escapes in a diffused manner. The surface treatment was carried out with a primer and subsequent coloring in light tones.

Liu Yang. China, 1998.

Her research strands include installation, graphic design, and curatorial methodology. She likes to experiment with mixed materials and explore new materials for public art. She participates in the activities of organizations such as the China Federation of Students and the Confucius Institute.

Fig. 8.1 a Vase table lamp
in the daylight. b Vase table
lamp in the darkness. c Vase
table lamp in the darkness.
Detail. Author: Luo Shihua,
pictures by Juri Ciani

Fig. 8.1   (continued)

8.3   Basalt Lamp

The work presented here is the result of a research project aimed at proposing basalt fiber as a suitable material for the realization of lamps. In fact, its dual nature allows it to be used as an opaque decorative material—which naturally generates its own reflections of light during the day (Fig. 8.3a, b)—and at the same time, when the lamp is switched on, as a technical filter that reduces the luminous intensity of the artificial light source, without altering its frequencies and thus keeping the chosen color alive (Figs. 8.4 and 8.5). The main volume is created with an elastic bandage filled with cotton, around which an RGB LED strip is wrapped; the whole unit is then enclosed in a cover, also made of cotton. A sheet of basalt fiber, impregnated with vinyl glue to increase its texture, was then cut into small surfaces and applied to cover the volume (Fig. 8.6). Three wooden sticks support and elevate the object. The result is that of a peculiar skin of the object (Manzini 1987), reminiscent of the scales of certain fish also because of the reflections of the basalt fiber scales, which change according to the source of light.

Nie Yurong. China, 1996.

Her research path, centered on visual art, is developed using different techniques such as acrylic paintings on canvas, realizations with everyday materials, and video art, with themes related to contemporary society. In the Florence area she participated at

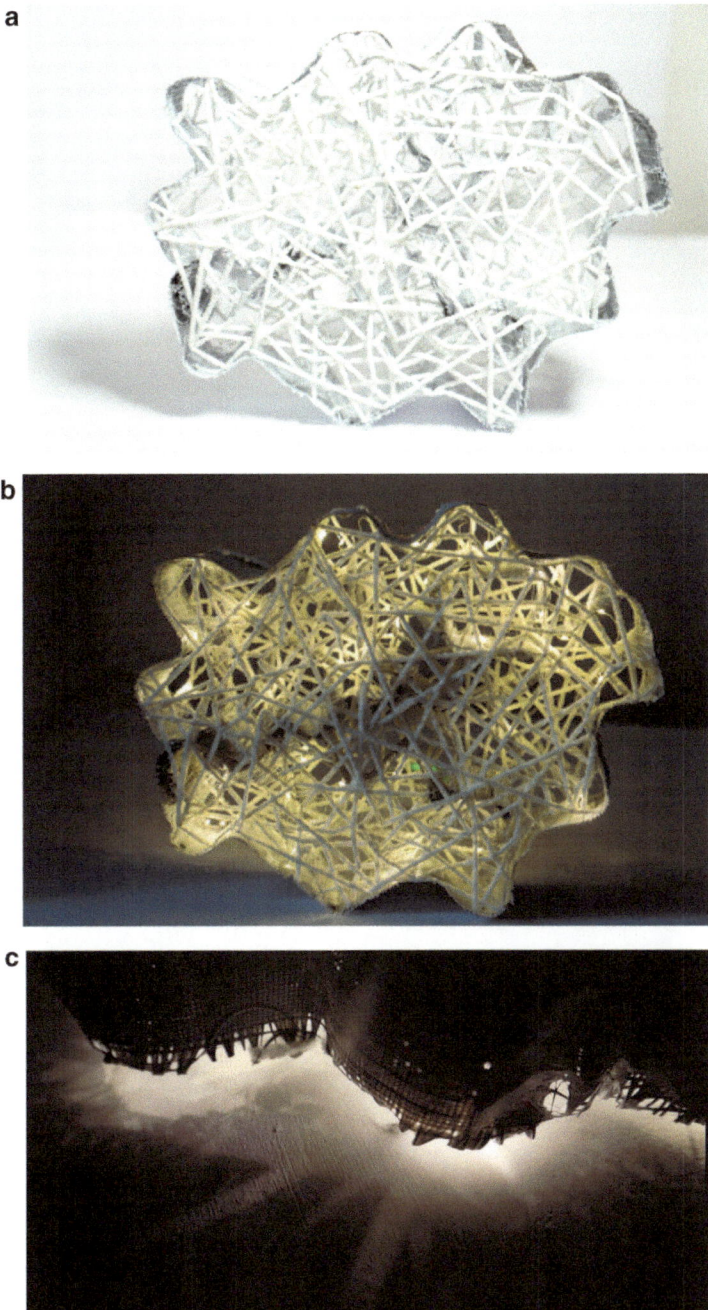

Fig. 8.2 **a** Sunbeam lamp in the daylight. **b** Sunbeam lamp in the darkness. **c** Sunbeam lamp in the darkness. Detail. Author: Liu Yang, pictures by Juri Ciani

Fig. 8.3 a Basalt lamp in the daylight. **b** Basalt lamp in the daylight. Detail of natural reflections of the basalt fiber. Author: Nie Yurong, pictures by Juri Ciani

"Mind the Map," a collective exhibition of students at Manifattura Tabacchi, "Hide and Seek" at PIA Palazzina Indiana Arte, "Perennial actuality" at Palazzo Strozzi, and the collective exhibition "Metamorphosis," in the art exhibition "Symbiosis."

Fig. 8.4   Basalt lamp in the darkness, with blue light. Author: Nie Yurong, picture by Juri Ciani

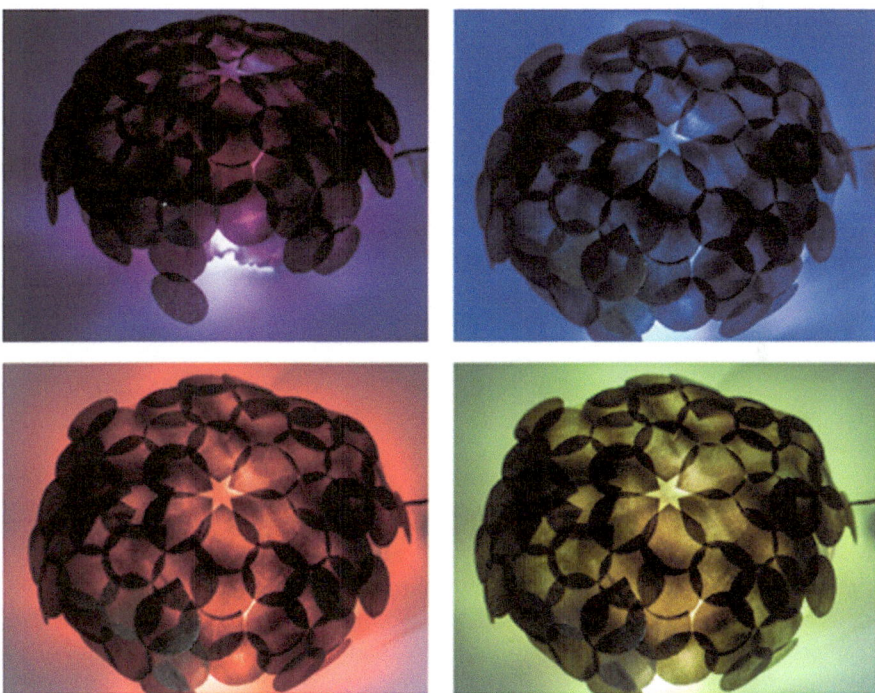

Fig. 8.5   Basalt lamp in the darkness, with several colors of light. Author: Nie Yurong, pictures by Juri Ciani

Fig. 8.6 Steps of lamp construction. Courtesy by Nie Yurong

8.4 Bas-abstract

Bas-abstract consists of an abstract composition painted directly onto woven basalt fiber in a flat weave. The purpose of this research was to test the qualities of the material, in particular by verifying how the basalt reacts to the application of acrylic paint and whether it can be used as a canvas on which to paint with traditional brushes. The positive outcome of the tests carried out led to the proposal to use the surface as a filter/diffuser: the properties of the material cause the painting to appear on the back of the fiber surface only when the light is on (Fig. 8.7), and then to disappear completely if the light is turned off (Fig. 8.8). This possibility of use also applies to technical areas, such as for lighting and signals in fireproof environments, with symbols to be selectively switched on only in case of emergency.

Fig. 8.7 Back of the painted basalt fiber surface, with light on. Author: Filippo Calmanti, picture by Juri Ciani

Fig. 8.8 Back of the painted basalt fiber surface, with light off. Author: Filippo Calmanti, picture by Juri Ciani

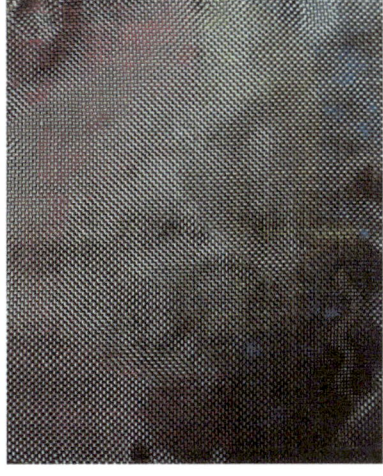

Filippo Calmanti. Macerata, 1997.

The research strand of his works focuses on the creation of environments that aim to modify the perception of space and the direct experience of the spectator, using mirror games, optical illusions, and color alterations. He has participated in "Building A New Word," 2019 National Arts Prize at the Accademia Albertina in Turin, "Open Art 2018-Piccoli apparati effimeri diffusi" at the Acccademia di Belle Arti in Florence, "Present and Future-just across the street" at MACRO Testaccio-Rome, 2017.

8.5   Meeting

This work (Fig. 8.9) is part of a study on the technical and expressive compatibility between different materials, both of which have a natural fibrous matrix, such as basalt fiber in the form of a flat-weave fabric, and small bamboo slats. The bamboo is used as a structural matrix (Fig. 8.10) onto which the basalt fiber is woven; in this case a white spray paint was used (Fig. 8.11). The lamp can be used for ambient lighting either on a desk or on a wall.

Zhu Liang. China 1996.

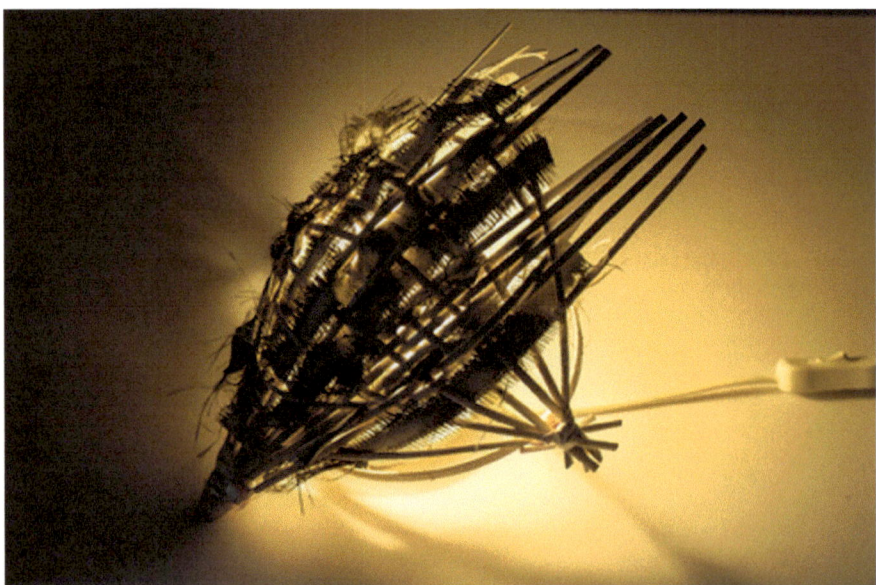

Fig. 8.9 Meeting lamp in the darkness. Author: Zhu Liang, picture by Juri Ciani

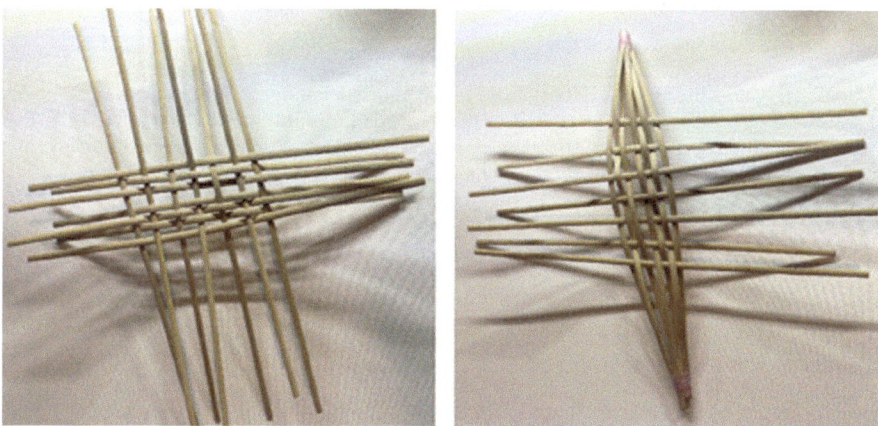

Fig. 8.10 *Meeting* lamp: weaving of bamboo sticks. Courtesy by Zhu Liang

Fig. 8.11 *Meeting* lamp. Detail of the white spray color over the basalt fiber surface. Author: Zhu Liang, picture by Juri Ciani

Her artistic research, inspired by the harmonic transformation of nature, focuses on urban public art. In her work, she often uses natural materials such as wood essences and in particular bamboo.

References

Castelli G, Antonelli P, Picchi F (eds) (2007) La fabbrica del design. Skira, Milano
Latouche S (2012) Bon pour la casse. Les déraisons de l'obsolescence programmèe. LES, Paris.
      Italian edition: Latouche S (2015) Usa e getta. Le follie dell'obsolescenza programmata (trans:
      Grillenzoni F) Bollati Boringhieri, Torino
Lee SM (1991) International encyclopedia of composites. VCH Publishing, New York
Manzini E (1987) La pelle degli oggetti. Ottagono 87:62–71
Turri MG (2011) Gli oggetti che popolano il mondo. Ontologia delle relazioni. Carocci, Roma

Chapter 9
Proposals, Out of the Box

This section presents a series of projects where basalt fiber is investigated, within art and design paths, in order to propose, understand and verify future uses not yet included in the material's range of applications, following the logic that "innovation is first and foremost a challenge to common sense and everyday attitudes (Jedlowski 2003: 116, cit. in Cerroni 2012: 25, our translation)." The properties of the fiber are tested in models in which the material is left exposed, and in which therefore the texture and appearance also play an important role in communicating the object and its use. Various materials have been combined and used in complementary ways with basalt fiber in order to optimize its performance and enhance its features. "During the early stages of the product development process, deliberate decisions about material use, energy efficiency, and waste avoidance can minimize or eliminate environmental impacts" (Ulrich et al. 2000: 233); for this reason, the inherent sustainability of the material was itself a source of inspiration for new types of product concepts.

9.1 Multiple

To obtain a sphere-like volumetric solid, a combination of regular hexagons and pentagons can be used, in a similar way to old soccer balls (five hexagons around each pentagon). Playing on this geometric rule, and adapting it to his own purposes, the designer has created a multifunctional table object/organizer, with a wooden frame and basalt fiber flat weave covering (Fig. 9.1a–c). Thus, the natural resistance of the fiber to abrasion is exploited, allowing for prolonged friction with different types of materials: in the spaces between one hexagon and the next, it is possible to insert objects of use, such as pens or other stationery products or smartphones, or other devices. The fiber, which here is not treated with resins, generates its own reflections of light that change the coloring on each face of the exploded solid. Form and mode of use are two levers of innovation, as defined by Rampino (2012: 41), applied in this case to product innovation.

Sun Yu Yao. China, 1996.

His artistic career is mainly centered on the use of installations, which he considers to be an intuitive and effective art form in order to "represent the whole range of emotions that an individual may feel and to explore the unknown, to investigate the uncertainties of life and all those questions that, especially after COVID-19 and the recent crisis, humanity asks itself on a daily basis." He participated in "Cabinet of Curiosities: Fantasy Museum," Shanghai, Duoyunxuan museum, 2018; "Imaginative Warehouse," Beijing, CHAO Museum, 2019; Exhibition of KOREA-CHINA Contemporary Design, Harbin, 2016; Exhibition "After," Florence, 2022.

9.2 Zen

Zen is a self-supporting shading module, presented here in a scaled model (Fig. 9.2). Taking advantage of basalt fiber's natural resistance to corrosion and, consequently, of its suitability for use in a marine environment, the project is a new typological proposal to complement sunshades, gazebos, and canopies. The main structure, in wood, is wrapped with textile elements in basalt fiber, which can be woven in different ways in order to control the quality of the generated shadow. The light reflections are always changing, depending on the direction of the sunlight. The introduction of basalt fiber in the context of the beach paves the way for possible experiments, for example as a sustainable material for beach equipment or furnishings, sunscreens—thanks to its solid shadows (Fig. 9.3)—or seasonal windbreaks.

Xia Xiaohaosheng. 1995, Sichuan, China.

His field of research concerns the sudden technological development of the twenty-first century and the consequent acceleration of the relationship with digitalization, which generates both greater ease of access to information but also a dependence of individuals on electronic devices.

9.3 Dream Hunter

In addition to the decorative meaning and spiritual symbolism of the dreamcatcher, the use of a material generated in the bowels of the Earth, untreated with resins or other substances, can express a deeper bond with the territory, also due to its own intrinsic sustainability. This product experimentation (Fig. 9.4a–c) utilizes the mechanical properties and weaving possibilities offered by the fiber (Fig. 9.5) and belongs to a field such as that of gift and souvenir items, where a greater attention to sustainability issues is undoubtedly desirable.

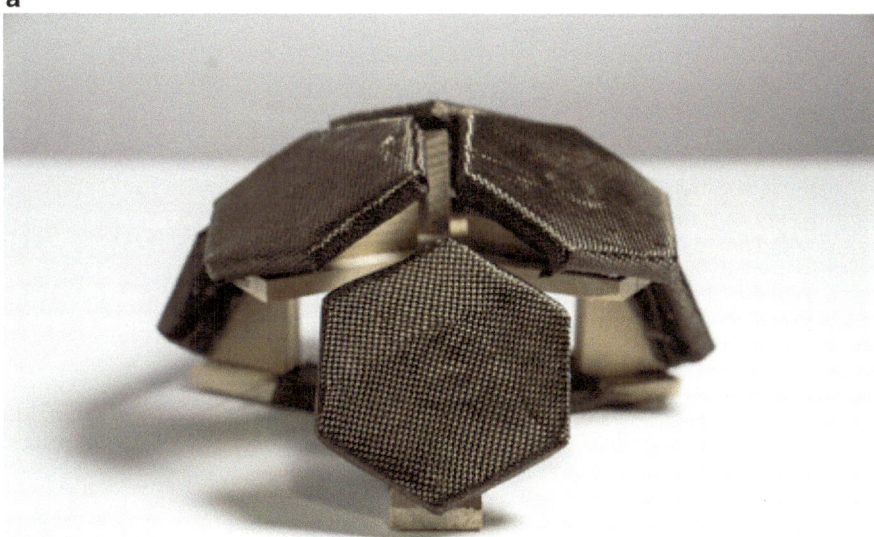

Fig. 9.1 a A multifunctional table organizer—sphere-like volumetric solid—obtained through a combination of regular hexagons. **b** Multifunctional table organizer in a different view. **c** Detail. Author: Sun Yu Yao, pictures by Juri Ciani

c

Fig. 9.1 (continued)

Fig. 9.2 *Zen*,
self-supporting shading
module. Author: Xia
Xiaohaosheng, pictures by
Juri Ciani

Fig. 9.3 The compact shade it generates and the weathering resistance of basalt fiber make *Zen* suitable in settings such as marine environments, bathing establishments, and generally for outdoor usage

Jiaqi Xu. Beijing, 1996.

Her research path is focused on understanding space and how contemporary art can fit into it. The subjects and content of her works stem from the experiences of everyday life, with the intention of stimulating reflection. In 2018 and 2019 she won two awards of excellence respectively in Japan and France, in 2021 she participated in the photo exhibition "Mind the Map" at the Accademia di Belle Arti in Florence, in 2022 she participated in the exhibition of the #dftm group of artists with her personal work *2050*.

9.4 Floating Shadow

This is an artistic investigation into the role and significance of temporary installations, summarized here by using a scaled model (Figs. 9.6 and 9.7). In the field of ephemeral architecture (Krauel 2010), materials that are not totally sustainable or not easily recyclable are often used, also because coverings or graphics are used that are not easy to separate. In this case, the attempt is to propose a model in which the walls play the role of both filter/containment and decoration/communication, using totally sustainable materials that are easy to dismantle and can also be reused separately.

Fig. 9.4 **a** Dream hunter, made with basalt fiber filaments. **b** Detail 1. **c** Detail 2. Author: Jiaqi Xu, pictures by Juri Ciani

a

Fig. 9.4 (continued)

Fig. 9.5 Continuous filaments of basalt fiber and its interweaving. Courtesy by Jiaqi Xu

The proposed elements are a wooden structure, basalt fiber tie-rods, and cladding/printing in sandwich panels of cork and wood.

Zhang Keqi. Shanxi, China, 1993.

In developing his works, he prefers neon as his main medium of expression. The artist focuses his attention on analyzing and codifying the contemporary world with a language influenced by spiritual thought, exploring the social relationships between individual and individual, individual and nature, and individual and society. He has exhibited his works at the Gattopardo Biennial (Palma di Montechiaro), at the Beijing Contemporary Art Exhibition, and at the second edition of "Essere," International Youth Art and Design Exhibition.

Fig. 9.6 Floating shadow, temporary installation with wooden structure, basalt fiber tie-rods, and cladding/printing in sandwich panels of cork and wood. Author: Zhang Keqi, pictures by Juri Ciani

Fig. 9.7 Floating shadow, temporary installation. Light effects in the darkness. Courtesy by Zhang Keqi

References

Cerroni A (2012) Il futuro oggi. Immaginazione sociologica e innovazione: una mappa fra miti antichi e moderni. Franco Angeli, Milano
Jedlowski P (2003) I fogli nella valigia. Sociologia, cultura, vita quotidiana. Il Mulino, Bologna
Krauel J (2010) Architettura effimera. Links books, Barcellona
Rampino L (2012) Dare forma e senso ai prodotti. Franco Angeli, Milano
Ulrich KT, Eppinger SD (2000) Product design and development. McGraw-Hill, Boston